VOLUME ONE HUNDRED AND THIRTY NINE

PROGRESS IN MOLECULAR BIOLOGY AND TRANSLATIONAL SCIENCE

Nanotechnology Tools for the Study of RNA

VOLUME ONE HUNDRED AND THIRTY NINE

PROGRESS IN MOLECULAR BIOLOGY AND TRANSLATIONAL SCIENCE

Nanotechnology Tools for the Study of RNA

Edited by

SATOKO YOSHIZAWA

Institute for Integrative Biology of the Cell (I2BC), CEA, CNRS, Univ Paris-Sud, Université Paris-Saclay, France.

ELSEVIER

AMSTERDAM • BOSTON • HEIDELBERG • LONDON
NEW YORK • OXFORD • PARIS • SAN DIEGO
SAN FRANCISCO • SINGAPORE • SYDNEY • TOKYO
Academic Press is an imprint of Elsevier

Academic Press is an imprint of Elsevier
50 Hampshire Street, 5th Floor, Cambridge, MA 02139, USA
525 B Street, Suite 1800, San Diego, CA 92101-4495, USA
125 London Wall, London EC2Y 5AS, UK
The Boulevard, Langford Lane, Kidlington, Oxford OX5 1GB, UK

First edition 2016

ISBN: 978-0-12-804565-7
ISSN: 1877-1173

For information on all Academic Press publications
visit our website at http://store.elsevier.com/

CONTENTS

CONTRIBUTORS

Spencer Carson
Department of Physics, Northeastern University, Boston, Massachusetts, USA

Robert Y. Henley
Department of Physics, Northeastern University, Boston, Massachusetts, USA

Takeya Masubuchi
Graduate School of Frontier Science, The University of Tokyo, Chiba, Japan

Hirohisa Ohno
Center for iPS Cell Research and Application, Kyoto University, Kyoto, Japan

Joseph D. Puglisi
Department of Structural Biology, Stanford University School of Medicine, Stanford, California, USA; Stanford Magnetic Resonance Laboratory, Stanford University School of Medicine, Stanford, California, USA

Hisashi Tadakuma
Institute for Integrated Cell-Material Sciences, Kyoto University, Kyoto, Japan; Graduate School of Frontier Science, The University of Tokyo, Chiba, Japan

Albert Tsai
Department of Applied Physics, Stanford University, Stanford, California, USA; Department of Structural Biology, Stanford University School of Medicine, Stanford, California, USA; Janelia Research Campus, Howard Hughes Medical Institute, Ashburn, Virginia, USA

Takuya Ueda
Graduate School of Frontier Science, The University of Tokyo, Chiba, Japan

Sotaro Uemura
Department of Structural Biology, Stanford University School of Medicine, Stanford, California, USA; Department of Biological Sciences, Graduate School of Science, The University of Tokyo, Bunkyo-ku, Tokyo, Japan

Meni Wanunu
Department of Physics, Northeastern University, Boston, Massachusetts, USA; Department of Chemistry and Chemical Biology, Northeastern University, Boston, Massachusetts, USA

Kazunori Watanabe
Department of Medical Bioengineering, Graduate School of Natural Science and Technology, Okayama University, Okayama, Japan

Koen Visscher
Departments of Physics and Molecular & Cellular Biology, College of Optical Science, The University of Arizona, Tucson, Arizona, USA

Takashi Ohtsuki
Department of Medical Bioengineering, Graduate School of Natural Science and Technology, Okayama University, Okayama, Japan

Hirohide Saito
Center for iPS Cell Research and Application, Kyoto University, Kyoto, Japan

PREFACE

Multifaceted roles that RNAs play in the cell constantly impose a technical challenge to those who study their functions and structures. RNAs, like other biological systems are in nanoscopic scale. Meanwhile, the remarkable progress in technologies in microfabrication has enabled manufacturing and assembling materials in nanometer scales as well as manipulating nano-objects. This advance has allowed the application of nanotechnology to manipulate or analyze directly individual biomolecules. RNAs are not exception. The power of nanotechnology has now been exploited in analyzes of RNA molecules.

This volume is devoted to pioneering works that represent integration of nanotechnology to RNA research. Application of nanotechnology pushes single molecule analysis of RNA one step forward. Nanophotonic structures called zero-mode waveguides (ZMWs) can reduce the volume necessary for an observation by more than three orders of magnitude relative to confocal fluorescence microscopy (down to the zeptoliter range) and allows single molecule observation at biologically relevant conditions (Chapter 1). Valuable biophysical properties can be characterized by applying mechanical forces to individual RNA molecules or using nanopores (Chapters 2 and 3). RNA can also be used as an element to form nanomaterials by conjugating to nanoparticles (Chapter 4). DNA, RNA itself or RNA with RNA binding protein can also form nanostructures and these nucleic-acid nanostructures can then be used as a support to exhibit biomolecules in a controlled geometry (Chapters 5 and 6).

I would like to express my sincere gratitude to the authors for their tremendous contribution. I would like to thank Dr. Michael Conn, Chief Editor of the Progress in Molecular Biology and Translational Science series for his initiative to have this volume in the series. I am grateful to Mary Ann Zimmerman, Senior Acquisition Editor, Helene Kabes, Senior Editorial Project Manager and Magesh Kumar Mahalingam, Project Manager at Elsevier for their continuous support. I hope that the readers of this volume will find its content useful and give them opportunities to think about how to incorporate these emerging new technologies into their own research.

Satoko Yoshizawa

CHAPTER ONE

Probing the Translation Dynamics of Ribosomes Using Zero-Mode Waveguides

Albert Tsai[*,†,‡], **Joseph D. Puglisi**[†,§], **Sotaro Uemura**[†,¶,1]

[*] Department of Applied Physics, Stanford University, Stanford, California, USA
[†] Department of Structural Biology, Stanford University School of Medicine, Stanford, California, USA
[‡] Janelia Research Campus, Howard Hughes Medical Institute, Ashburn, Virginia, USA
[§] Stanford Magnetic Resonance Laboratory, Stanford University School of Medicine, Stanford, California, USA
[¶] Department of Biological Sciences, Graduate School of Science, The University of Tokyo, Bunkyo-ku, Tokyo, Japan
[1] Corresponding author: e-mail address: uemura@bs.s.u-tokyo.ac.jp.

Contents

Progress in Molecular Biology and Translational Science, Volume 139
ISSN 1877-1173
http://dx.doi.org/10.1016/bs.pmbts.2015.10.006

Abstract

In order to coordinate the complex biochemical and structural feat of converting triple-nucleotide codons into their corresponding amino acids, the ribosome must physically manipulate numerous macromolecules including the mRNA, tRNAs, and numerous translation factors. The ribosome choreographs binding, dissociation, physical movements, and structural rearrangements so that they synergistically harness the energy from biochemical processes, including numerous GTP hydrolysis steps and peptide bond formation. Due to the dynamic and complex nature of translation, the large cast of ligands involved, and the large number of possible configurations, tracking the global time evolution or dynamics of the ribosome complex in translation has proven to be challenging for bulk methods. Conventional single-molecule fluorescence experiments on the other hand require low concentrations of fluorescent ligands to reduce background noise. The significantly reduced bimolecular association rates under those conditions limit the number of steps that can be observed within the time window available to a fluorophore. The advent of zero-mode waveguide (ZMW) technology has allowed the study of translation at near-physiological concentrations of labeled ligands, moving single-molecule fluorescence microscopy beyond focused model systems into studying the global dynamics of translation in realistic setups. This chapter reviews the recent works using the ZMW technology to dissect the mechanism of translation initiation and elongation in prokaryotes, including complex processes such as translational stalling and frameshifting. Given the success of the technology, similarly complex biological processes could be studied in near-physiological conditions with the controllability of conventional in vitro experiments.

1. INTRODUCTION

Within cells, proteins perform the bulk of the biochemical, structural, and regulatory activities required to maintain life. However, the genes that code for these proteins are composed of nucleic acids; they must be translated into the proper sequence of amino acids using an adaptor molecule, the transfer RNAs (tRNAs). The ribosome, a multimega Dalton complex with a functional core composed of ribonucleic acids [ribosomal RNA (rRNA)] with numerous peripheral proteins, is the central catalytic machinery that ensures an optimal balance between selecting for the correct tRNA and the speed at which nascent peptides are synthesized. Translation is energetically intensive, both in terms of the GTPs consumed directly during the process and the ATPs to aminoacylated tRNAs[1,2] and indirectly to synthesize and maintain the translation machinery.[3,4] Therefore the ribosome must correctly coordinate its interactions with translation factors, tRNAs, and messenger RNAs (mRNAs) to ensure that protein synthesis is efficient, specific, and well regulated.

Because translation is the crucial final step in expression of genetic information, the process has been under intense study ever since the ribosome was identified as the molecule catalyzing translation more than half a century ago.[5–9] Numerous biochemical studies have measured the kinetic rate of tRNA selection and rejection, peptide bond formation, and translation factor binding.[10–14] Structural studies using X-ray diffraction and cryoelectron microscopy have resolved the architecture of the ribosome[15–17] and captured key intermediates with the relevant tRNAs and translation factors along the translation pathway,[18–21] illustrating the physical mechanisms behind these processes. As a result, extensive information is available concerning the kinetics and the relevant structures for each individual step along the translation pathway.

2. THE RIBOSOME MUST CHOREOGRAPH COMPLEX INTERACTIONS BETWEEN TRANSLATION FACTORS, tRNAs, AND mRNAs

More than simply a static collection of individual structural states and biochemical steps, translation is a dynamic process, where these states and steps are linked together through complex interactions. The translation machinery must transition from one structural state into another in a coherent and seamless manner; the transitions are frequently triggered by specific biochemical changes. For translation to proceed efficiently and with high fidelity, the ribosome must synergistically coordinate these compositional, conformational, and biochemical changes so that they occur in the correct sequence. With the large number of players involved, including the mRNA, tRNAs, translation factors, and various parts of the ribosome itself, the possible pathways for any given process become immense. Thus understanding how the ribosome and other molecules evolve between critical structural states and which biochemical steps drive or result from these structural rearrangements is central to outlining how the individual pieces fit together coherently in the global dynamics of translation.

In theory, experiments to gain such information would be straightforward, involving assays that could directly track the ribosome and translation factors across multiple steps during translation. In practice, such experiments are difficult to conduct as crucial parts of the process involve multiple stochastic structural rearrangements and biochemical reactions in rapid succession. Moreover the intermediates of those steps could additionally contain heterogeneous populations of ribosomes loaded with different translational factors and in different conformations. The stochastic and linked nature of these steps renders the global dynamics of translation difficult to track using bulk techniques where synchronizing molecules is difficult and the large number of molecules mask heterogeneous populations.

The advent of single-molecule techniques to probe biological systems, ranging from optical tweezers to fluorescence microscopy, has made important inroads into answering some key questions concerning the dynamics of translation.[22,23] With their ability to track individual molecules directly, they provide an answer to the challenges of measuring stochastic processes with potentially heterogeneous subpopulations exhibiting different behaviors. Using fluorescence microscopy, both the composition and conformation of the translational machinery can be tracked directly with labeled components through multiple stochastic steps. The ability to distinguish individual ribosomes with different translation factors additionally allows for the behaviors of different subpopulation to be separated when the results are analyzed. The presence of each subunit of the ribosome and their conformational state could be monitored through labeling the two subunits of the ribosome at specific locations.[24–26] Additionally, tRNAs[27] and protein translation factors[28] could be monitored by labeling them using a variety of techniques. Accordingly, multiple studies have taken advantage of fluorescence microscopy to probe the dynamics of the ribosome during all phases of translation. These studies have shown that the ribosome functions through coordinating a core set of conformational changes linked to the binding, dissociation, and structural rearrangements of tRNAs and translation factors. Specifically, these changes involve spontaneous local structural rearrangements around the L1 stalk of the prokaryotic ribosome near the deacylated tRNA exit site (E site) and a global intersubunit rotation remodeling numerous contacts along the subunit interface that is driven by the energy of either GTP hydrolysis or peptide bond formation.[28–31] Moreover, ribosomes also monitor and manipulate the conformations of the tRNAs to select the correct tRNA for accommodation, as well as catalyzing the GTPase activities of several translation factors to trigger critical structural rearrangements that drives the translation pathway forward.[32–34]

3. THE CHALLENGES OF OBSERVING COMPONENTS OF THE TRANSLATION MACHINERY AT HIGH CONCENTRATIONS

In single-molecule experiments, the challenges of reconstituting the translation machinery in vitro frequently required the use of simplified model systems designed to probe particular aspects of translation. These simplified model systems provided focused information on the dynamics of the ribosome along well-defined pathways at specific points in the translation cycle; however, working with mRNAs coding for natural sequences at near-physiological conditions remained a significant challenge. Fundamentally, the technical barriers are twofold: (1) the need to track multiple components of the translation machinery at once to correlate the actions of several different players and (2) the need to work at sufficiently high concentrations of labeled ligands to observe an extended process within a finite amount of time before the fluorophores photobleach, permanently eliminating their fluorescence. The first criterion requires microscopes that could simultaneously monitor multiple-labeled ligands labeled using different fluorophores that are separated using their emission spectra. While the optics requires careful engineering, fluorescence microscopes with multiple laser lines for exciting different fluorophores and the emission detection channels to detect them are available for this purpose.

The barrier posed by increasing the concentrations of labeled molecules in solution is significantly more challenging from a technical perspective. The fundamental problem that prevents increasing the concentration of the molecule of interest to their in vivo conditions stems from the background fluorescence that freely diffusing fluorophores generate[23] (Fig. 1A). At low concentrations, a bound molecule labeled with a fluorophore anchored on the surface of a quartz microscope slide[35] would be brighter than molecules diffusing in solution as its fluorescence occurs within a diffraction-limited volume versus occurring throughout the volume in which it diffuses for a given amount of time. Nevertheless, at sufficiently high concentrations of diffusing molecules, the background fluorescence becomes sufficiently high to result in poor signal-to-noise

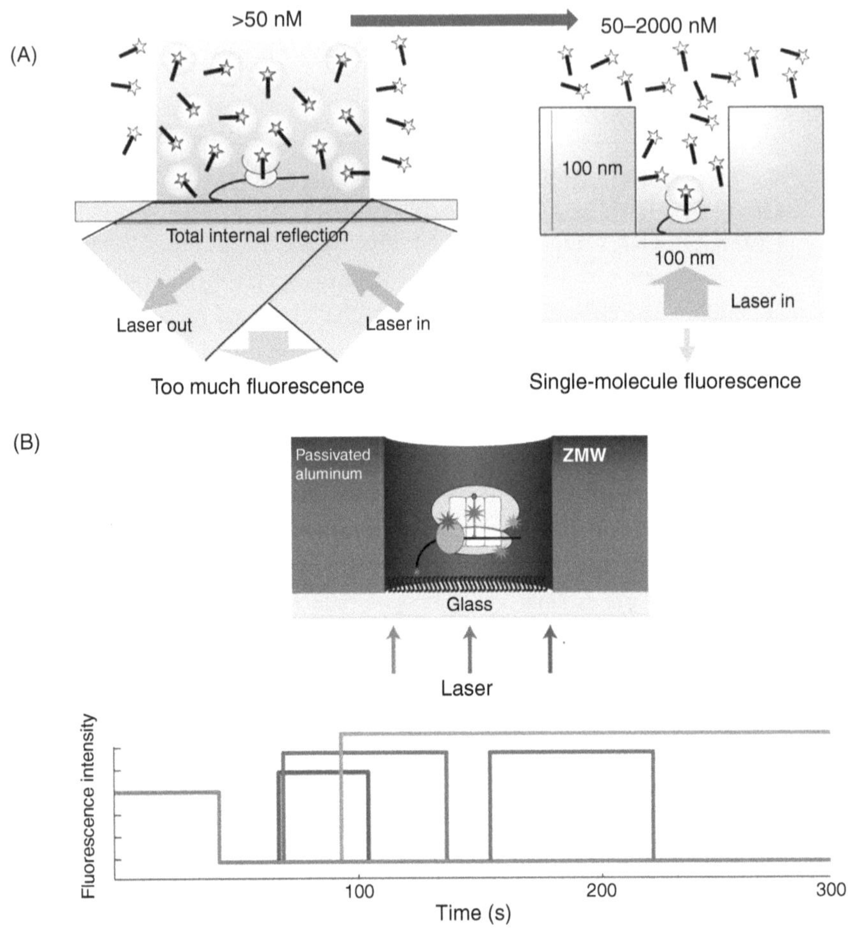

Figure 1 Zero-mode waveguide (ZMW) allows translation to be tracked using single-molecule fluorescence at near-physiological conditions. (A) Conventional single-molecule setup using total internal reflection fluorescence (TIRF) microscopy illuminates about 150 nm into the solution containing freely diffusing molecules labeled with fluorophores. This limits the concentration of labeled ligands in solution down to 50 nM or less at normal imaging speed of 10 times per second. ZMWs concentrate the laser excitation down to only 20–30 nm from the bottom of the well, allowing more than 1 μM of labeled molecules to be present in solution. (B) In order to make the ZMW structure compatible with biological molecules, the aluminum surface was passivated and the exposed glass surface at the bottom of the ZMW was treated with biotinylated polyethylene glycol (PEG). Biotinylated molecules can be immobilized inside the ZMW through the use of streptavidin or neutravidin. The increased concentration and the multiplexing ability of the ZMW microscope to observe several different color channels simultaneously allowed the dynamics of ribosomes translating on complex mRNA sequences to be tracked with a full complement of translation factors and tRNA at near-physiological concentrations.

ratios, eventually drowning out the signal from the bound molecule altogether. Extending the camera exposure time would average out this background; however, an exposure time longer than a 100 ms would begin to miss most of the critical dynamics of the ribosome. Therefore the rates of the individual steps of translation would limit the maximum camera exposure time to a few hundred milliseconds. Working with the optics of microscopes, various techniques can effectively restrict the laser illumination to only excite bound fluorophores while those freely diffusing are not illuminated. A common approach is to angle the laser beam shallower than the critical angle so that the beam is totally reflected back into the glass slide.[36] The internally reflected laser beam creates an exponentially decaying evanescent wave that only penetrates about 150 nm into the solution above the glass slide. TIRF microscopy can thus tolerate up to 50 nM of free labeled molecules when they are imaged at a rate of 10 times per second (100 ms of exposure time per frame),[37] an improvement of over a factor of 10 compared to direct illumination. However, the biological concentrations of translation factors and tRNAs are usually in the hundreds (nanomolar to micromolar range),[38–40] requiring an additional improvement by around a factor of 50. With ligands at tens of nanomolar, the waiting time until tRNAs or translation factors bind to the ribosome would extend from 10 to 100 ms in vivo to more than a few seconds. Excessively long waiting times would interrupt the process being observed and significantly increase the likelihood of off-pathway reactions that are nonphysiological and detrimental. Furthermore, long waiting times limit the ability to track multistep processes in their entirety because of the finite lifetime of a fluorophore before it photobleaches—irreversibily reacting such that fluorescence emission is destroyed. This loss of signal weakens a key asset of single-molecule experiments to track individually the components of a system continuously in time.

4. ZERO-MODE WAVEGUIDE FLUORESCENCE MICROSCOPY ALLOWS THE TRANSLATION MACHINERY TO BE TRACKED AT HIGH CONCENTRATIONS OF LABELED LIGANDS

Nanophotonic devices, due to their ability to create optical effects in a highly localized and nonlinear manner, provide a solution to further reduce unnecessary illumination of labeled molecules in solution while concentrating the excitation laser to the bound molecule of interest.

ZMWs are one such nanostructure that can confine laser illumination to an extremely small region above a bound fluorescent molecule.[41,42] A ZMW is a well with a diameter of 100–150 nm etched into a thin 100 nm layer of metal, usually aluminum, deposited on the surface of quartz. As each ZMW well is smaller than the wavelength of visible light, laser illumination in the range of 700–400 nm commonly used in fluorescence spectroscopy cannot propagate through them. The metal walls strongly quench the illumination, leaving only the first 10–30 nm above the surface illuminated, depending on the diameter of the ZMW. Combined with the small dimensions of the well itself, the illumination volume within each ZMW is limited to be on the order of zeptoliters (10^{-21} L). This extremely small illumination volume could in theory allow the presence of more than several micrometers of freely diffusing fluorescent molecules while maintaining acceptable signal-to-noise ratios to detect a single fluorophore bound in the bottom of the ZMW (Fig. 1B). The higher concentration limit now encompasses the average equilibrium dissociation constant (K_d) of specific ligands for a significant portion of enzymes,[43] allowing their dynamics to be tracked at near-physiological conditions.

ZMW-based single-molecule fluorescence microscopy provides a powerful tool that is a good match for measuring the dynamics of the translation machinery. However, the nanofabrication process to produce thousands of geometrically consistent ZMW wells,[44] the surface chemistry to passivize the metal surface to be compatible with proteins and nucleic acids,[45] and the optical expertise to image from ZMW chips[46] present serious hurdles to a general adoption of ZMW-based microscopy for use in probing biological processes. Fortunately, Pacific Biosciences has developed their next generation sequencing system, the RS, using the ZMW technology to image fluorescent nucleotides binding to single polymerases,[47,48] providing a commercially available platform that has been developed for biological applications. This technical foundation opens the door to using ZMW fluorescence microscopy to study translation at near-physiological contexts. This chapter will provide a review of the recent developments in adapting ZMW fluorescence microscopy to study the global dynamics of translation in increasingly realistic and complex setups and highlight the key findings of these studies. The mechanistic insights garnered from these studies showcase the potential utility of the ZMW technology as a single-molecule fluorescence technique to probe complex biological processes.

5. TRACKING tRNA TRANSITIONING THROUGH ELONGATING RIBOSOMES INSIDE ZMWs AT NEAR-PHYSIOLOGICAL CONDITIONS

The primary advantage of ZMWs to suppress background fluorescence from labeled molecules in solution is especially valuable in experiments where high ligand concentrations are needed to promote binding in a timely manner. The initial effort in employing ZMWs to probe translation focused on the repetitive binding and transit dynamics of tRNAs through the ribosome during elongation. During this phase, tRNAs serve as the adaptor molecule that carries into the ribosome the appropriate amino acid that is encoded by an mRNA codon. A new tRNA must bind to the ribosome for each codon on the mRNA, necessitating the removal of the tRNA used for the previous codon as well. Through each cycle of elongation, the ribosome undergoes a coordinated set of structural (conformational) changes and biochemical steps to select the correct tRNA to be accommodated in the A site, add the amino acid of the new tRNA onto the nascent peptide, and then eject the deacylated old tRNA through the E site. During its time inside the ribosome, a tRNA must move from the aminoacyl-tRNA site (A site) into the peptidyl-tRNA site (P site) through translocation, a process requiring the GTPase elongation factor G (EF-G) as a catalyst to proceed efficiently. Before the tRNA can depart from the ribosome, a subsequent round of tRNA accommodation followed by translocation must move the tRNA from the P site into the E site. However, the timing of E-site tRNA ejection was ambiguous—evidence for spontaneous tRNA dissociation once the ribosome translocates or controlled ejection of the E-site tRNA contingent upon the arrival of a new tRNA in the A site both surfaced.[49–52] A direct approach to address this problem using single-molecule fluorescence would be to track two or more different labeled tRNAs transiting through ribosomes across multiple rounds of elongation and observe how the signal for the A-site tRNA correlates with that of the E-site tRNA. While labeled tRNAs had been used in single-molecule experiments, the labeled tRNA concentrations had to be kept below 50 nM to limit background fluorescence in TIRF microscopy. The waiting times for labeled tRNAs to accommodate in the A site alone were in excess of 10 s in some cases; the tRNAs would photobleach before they can reach the E site, which additionally requires one more tRNA binding step and two translocation steps.[27,29,53]

As a result, these studies concentrated on specific steps of the elongation cycle, measuring the dynamics of tRNA accommodation or conformational sampling of the tRNA immediately surrounding translocation. In order to answer if E-site tRNA dissociation is linked to A-site tRNA arrival using labeled tRNAs, starting with a ribosome preloaded with a tRNA in the P site, at least one elongation cycle must occur with labeled tRNAs to move a tRNA into the E site. The arrival of the A-site tRNA can then be correlated to the dissociation of the E-site tRNA. The concentration of labeled tRNAs in solution must be significantly higher than TIRF conditions so that the combined time for an entire elongation cycle and then for a tRNA to bind to the A site must be short enough to ensure that the disappearance of the signal from the E-site tRNA is due to dissociation instead of photobleaching.

6. SURFACE INACTIVATION PREVENTS PROTEIN AND NUCLEIC ACID AGGREGATION ON METAL SURFACES

The requirement for conducting elongation experiment at high concentrations of labeled tRNAs highlights the benefits of performing single-molecule tRNA transit experiments using ZMW microscopy. The ability of ZMW fluorescence microscopy to limit unnecessary laser illumination beyond 30 nm above the quartz surface within a waveguide critically depends on the metal sidewalls, which quench the laser illumination by virtue of being conductive. Unfortunately, metal surfaces attract charged molecules such as nucleic acids and proteins with charged amino acids. If left unchecked, this attraction could lead to significant aggregation or nonspecific attachments of molecules to the metal surfaces and to each other. This would be especially problematic if labeled molecules are involved, creating background fluorescence that could completely mask the signal from a single immobilized ribosome at the bottom of a ZMW. Even with unlabeled molecules aggregating, a crowded surface could change the dynamics of any biological molecule through nonspecific and specific interactions with the molecules in solution. In severe cases, the concentration of the sticking molecule in solution and at the surface would also be significantly skewed. Therefore a variety of surface treatments have been devised to passivate the metal surfaces for use in biological applications. Specifically, Korlach et al. demonstrated that derivatizing the metal surface with poly(vinylphosphonic) acid significantly limits nucleic acid and protein attachment to aluminum surfaces[45] (Fig. 1B). Before conducting single-molecule tRNA transit experiments with immobilized ribosomes, Uemura et al. verified that the surface treatments are sufficient to prevent ribosomes, tRNAs, and translation factors from sticking to the metal surfaces.[54] By additionally treating the quartz surface on the bottom of ZMWs with biotinylated polyethylene glycol (PEG), a chip containing thousands of ZMW wells could be used to observe individual ribosomes attached via biotin–streptavidin (or substitutes such as neutravidin) interactions, either directly or through the mRNA.

7. TRACKING tRNA TRANSITING THROUGH THE RIBOSOME THROUGH MULTIPLE ROUNDS OF ELONGATION

Ribosomes immobilized in ZMWs are functionally active. With the small (30S) subunit of the ribosome attached to the bottom of ZMWs through a 5'-biotinylated mRNA, these subunits successfully formed 30S preinitiation complexes (PICs) when initiation factors 1, 2, and 3 (IF1, IF2, and IF3); the initiator tRNA (fMet-tRNA^{fMet}); and GTP were delivered to the ZMW chip. When the large (50S) subunit was delivered to these 30S PICs, 70S initiation complexes (ICs) ribosome were formed, demonstrating that complete ribosomes could be assembled through the canonical translation initiation pathway. tRNA–tRNA FRET experiments, using ICs formed with Cy3-labeled initiator tRNA and Cy5-labeled tRNA^{Phe} delivered in solution as a ternary complex with EF-Tu(GTP), verified that these ICs were functionally competent. On an mRNA coding for Met followed by Phe, the FRET signal evolved into a stable high-FRET state. Changing the codon from Phe to the near-cognate Leu gave the expected result of frequent and short tRNA sampling that remained in the low-FRET state. Adding tetracycline to the solution inhibited tRNA accommodation but not tRNA binding; therefore the tRNAs only proceeded up to an intermediate FRET state with the drug present. These observations align with previous experiments in bulk using single-molecule fluorescence.[27,34] Finally, when Cy5-labeled tRNA^{Phe} and Cy2-labeled tRNA^{Lys} were delivered to ribosomes on mRNAs coding for various combinations of Phe and Lys codons [eg, 6(FK) containing six repeating Phe-Lys or 4(FKK) containing four repeating Phe-Lys-Lys], the correct tRNA signal sequence was seen, demonstrating that ribosomes assembled inside of ZMW wells are fully functional through initiation and elongation [Fig. 2A showing 6(FK)]. With increasing

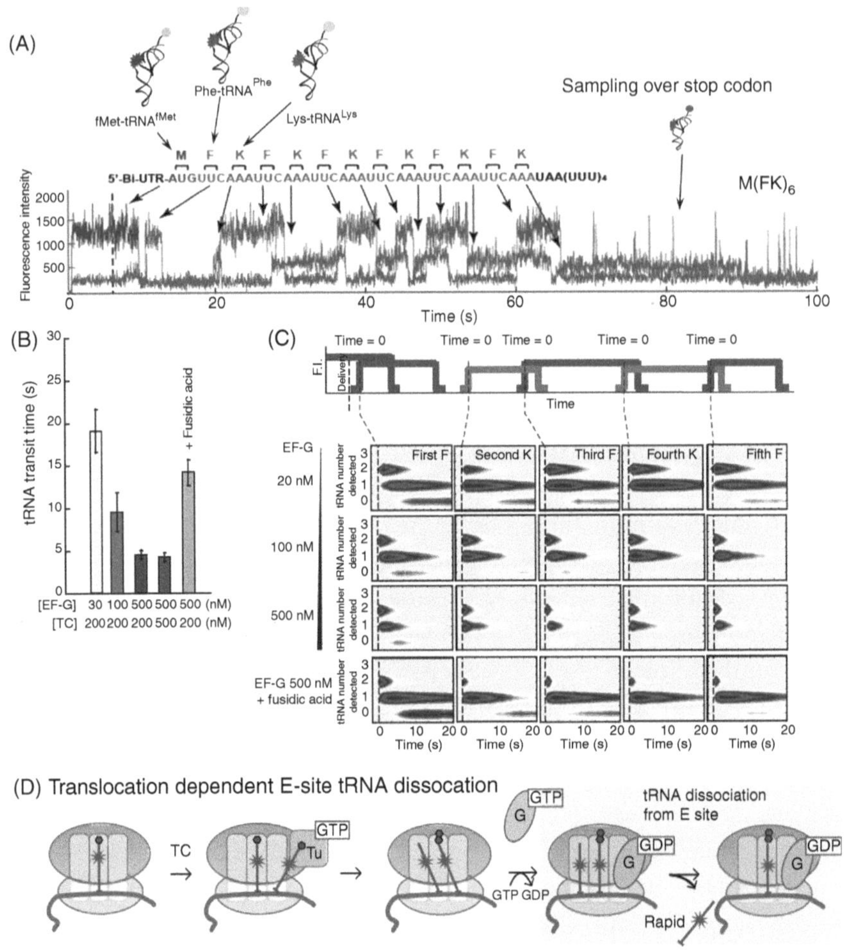

Figure 2 Applying the ZMW technology to study translation: tRNA transit through the ribosome during elongation. (A) Tracking tRNA transit through the ribosome during elongation at near-physiological concentrations of fluorescently labeled tRNAs (tRNA^{fMet} with Cy3, tRNA^{Phe} with Cy5, and tRNA^{Lys} with Cy2, where each tRNA is indicated using their one-letter amino acid code in the sequence of pulses) was possible through imaging ribosome complexes in ZMW wells. Translating an mRNA coding for six repeating Phe and Lys codons gave the correct sequence of fluorescent pulses (alternating F and K pulses), verifying that ribosomes translate normally within ZMWs. Many ribosomes translated the full message and begin to show short tRNA sampling pulses over the final stop codon as no termination and ribosome release factors were available to terminate translation. (B) Working at high concentrations of tRNA and EF-G, translation elongation was efficient and processive. With 500 nM of labeled tRNAs in solution, translation could proceed at a rate of less than 5 s per codon under single-molecule imaging conditions. (C) Tracking tRNA occupancy on the ribosome during translation elongation revealed that tRNA departure from the E site

concentrations of tRNA ternary complexes and EF-G (an elongation factor that catalyzes translocation), each cycle of elongation became progressively faster up to 3–4 s per round of elongation at 500 nM of labeled tRNA and EF-G (Fig. 2B). These are near-physiological concentrations that were previously unattainable using conventional TIRF microscopy. As tRNA dwell times on the ribosome did not change significantly when the tRNA ternary complex concentration was lowered to 200 nM (EF-G is still held at 500 nM); this suggests that the rate-limiting step in the current setup lies at translocation. At sufficiently high tRNA and EF-G concentrations, a significant fraction of ribosomes translated the full 12 codons of the 6(FK) mRNA before pausing over the stop codon because there were no termination factors in solution. Transient tRNA sampling pulses of approximately 50 ms each could be observed and their frequency increased with increasing tRNA concentration.

8. TRACKING tRNA TRANSIT AT HIGH CONCENTRATIONS REVEAL A STOCHASTIC tRNA EXIT MECHANISM FROM THE E SITE

In addition to verifying that ribosomes can translate within ZMWs, tRNA transit experiments tracking tRNA dynamics across multiple codons provide the necessary data to examine in detail the mechanism of E-site tRNA dissociation. Plotting the tRNA occupancy of each ribosome over time for each codon (Fig. 2C) revealed that ribosomes very rarely (<2% of the time) have all three of its tRNA sites occupied, even at high concentrations of both tRNAs and EF-G. On the other hand, ribosomes frequently had two tRNAs, presumably in the A and P site, as seen in the overlap of the tRNA signals. These results may tentatively support the hypothesis that the

◀ is dependent on translocation. As translocation rates increased with increasing EF-G concentration, the time that two tRNAs occupied the ribosome decreased. With fusidic acid blocking EF-G dissociation and thus A-site tRNA arrival, ribosomes still quickly evolved into single-tRNA occupancy. E-site tRNA departure therefore occurs stochastically after translocation, without clear correlation to the arrival of a new tRNA in the A site. (D) This result supports the model of E-site tRNA departure where translocation is the gating event immediately prior to tRNA dissociation. As soon as the ribosome translocates, the E-site tRNA rapidly and stochastically departs without the need for another tRNA to accommodate in the A site. *Adapted with permission from Uemura et al.*[54]

E-site affinity for tRNA decreases when a new tRNA occupies the A site. However, the duration of the overlap was proportional to the time needed to translocate the ribosome, as evidenced by its direct dependence on the concentration of EF-G, which catalyzes translocation with the energy of GTP hydrolysis. Additionally, ribosomes with two tRNAs quickly evolved into the single-tRNA state even when fusidic acid was added. Fusidic acid locks EF-G in its GDP-bound state, occluding only the arrival of a new tRNA to the A site but allows the E-site tRNA to freely dissociate. These results support the hypothesis that tRNA departure from the E site occurs spontaneously after translocation without any clear correlation to the arrival of a new A-site tRNA, favoring the model that has translocation as the event that gates E-site tRNA release (Fig. 2D). The work of Uemura et al. established the utility of the ZMW technology in studying the global dynamics of translation in its entirety at near-physiological conditions. It is possible now to monitor tRNA dynamics throughout the elongation process on a coding mRNA using ribosome complexes assembled using the canonical translation initiation pathway. With the higher attainable ligand concentrations, major processes of translation are observable within a timeframe of seconds to minutes, which is within the average photobleaching lifetimes of good organic fluorophores.

9. DISSECTING THE MECHANISM OF INITIATION AND ELONGATION

In addition to tRNAs, the ribosome also interacts with numerous translation factors during all phases of translation to ensure that the processes occur in a well regulated and efficient manner. During initiation of bacterial protein synthesis, several IFs guide the assembly of the two subunits of the ribosome into the correct conformation over the start codon of the mRNA, creating an elongation-competent ribosome in a regulated manner.[55] During elongation, EF-G uses the energy of GTP hydrolysis to promote translocation of the ribosome, which involves a host of local and global structural rearrangements both within the ribosome and on EF-G.[56] Therefore both initiation and translocation involve a succession of stochastic steps, controlled by translation factor binding or dissociation, in addition to conformational changes of the translation factors and the ribosome. The timing of binding, dissociation, and conformational changes is tightly choreographed to ensure that these steps occur synergistically without interfering with each other. Bulk biochemical assays provide valuable information on the overall outline of the steps in initiation[57,58] and translocation,[59–61] as well as identifying the functions of the translation factors; however, their results suggested different pathways with different ordering of events, which became areas of disagreements and debate; that some ribosomes proceed through alternative pathways also remained a distinct possibility, one that is difficult to address with techniques that measure the aggregate behavior of many molecules (eg, a picomole, a common amount used in biochemical assays, contains more than 10^{11} molecules). Direct measurements of ribosomes undergoing initiation and translocation would require tracking the ribosomes along with multiple ligands for times up to several minutes. Similar to the tRNA transit experiments, these experiments greatly benefit by working at high concentrations of labeled ligands in solution to expedite all the necessary steps while decreasing the risk of off-pathway reactions. As a result, ZMW technology is well situated to dissecting the mechanism of translation with multiple interconnected compositional and conformational changes.

10. DEFINING THE PATHWAY TO ASSEMBLING A PREINITIATION COMPLEX AND TRANSITIONING INTO ELONGATION

When a molecule of mRNA is available for translation, only the small subunit of the ribosome binds initially to the mRNA. In prokaryotes, the Shine–Dalgarno sequence on the mRNA situates the 30S subunit approximately over the start codon.[62,63] Therefore the goal of initiation is to determine if translation should proceed, ensure the correct reading frame for the translated protein, and then guide the assembly of the large subunit to the small subunit to form an elongation competent 70S complex. The first major intermediate step of initiation is to form a 30S PIC[64,65] composed minimally of the 30S subunit, mRNA, the initiator tRNA ($tRNA^{fMet}$), and the GTPase IF2. IF2 is a critical protein that lies on the intersubunit interface, guiding the two subunits into a specific conformation after subunit joining[28,66–69] and stabilizing the initiator tRNA in the P site.[70–72] The formation of the 30S PIC ensures that the ribosome is in the correct reading frame on an mRNA[73] by placing the start codon in the P site of the small subunit using the codon–anticodon interaction between the mRNA and the initiator tRNA. Two more IFs in prokaryotes, IF1 and IF3,[20] also participate in this process by preventing premature subunit association and enhancing selection for an initiator tRNA in the P site.[74–78] The 30S PIC serves as the platform for subunit joining to occur, yet the pathway leading up to its formation has been debated, particularly concerning the exact order in which the initiator tRNA and IF2 binds to the 30S. Results from different studies suggested sequential binding of IF2 and the initiator tRNA based on measured kinetic binding rates,[79] while additional evidence pointed to the two molecules forming a complex first in solution.[64,80] After 50S subunits join with the 30S PIC, several events must occur within a short time frame ending with an elongation competent 70S complex: GTP hydrolysis by IF2, IF2 dissociation from the ribosome, and the arrival of the first elongator tRNA as dictated by the 2nd codon on the mRNA. These steps end the initiation phase of translation and transition the ribosome into the elongation phase, breaking the interaction between the ribosome and the Shine–Dalgarno sequence of the mRNA in the process.[81,82] Ordering the events in initiation, which occur in rapid succession, proved to be challenging without the ability to track individual ribosomes. The final gating event that controls the arrival of the elongator tRNA was a debated question with either IF2 departure or GTP hydrolysis as the top candidates.

In order to clarify the mechanism of the formation of the 30S PIC, Tsai et al.[83] performed experiments with all of the essential components of the late initiation pathway fluorescently labeled, taking advantage of the ability of a ZMW microscope to observe simultaneously multiple color channels (Fig. 3A)—the 30S subunit with Alexa488 (30S-A488), the initiator tRNA with Cy3 ($\text{tRNA(Cy3)}^{\text{fMet}}$), IF2 with Cy5 (IF2-Cy5), the 50S subunit with Cy3.5 (50S-Cy3.5), and the elongator tRNA with Cy2 ($\text{tRNA(Cy2)}^{\text{Phe}}$). At low concentrations of the initiator tRNA and IF2, sequential arrival was the dominant pathway to assemble the 30S PIC (Fig. 3B). The exact order depended on the concentrations of not only the two molecules themselves, but also on the presence of IF1 and IF3. This is consistent with the role of these two nonenzymatic factors to modulate the affinity of IF2 and the initiator tRNA to the ribosome.[68,72,84] With increasing concentrations of $\text{tRNA}^{\text{fMet}}$ and IF2, both molecules arrived simultaneously to the 30S subunit with increasingly higher proportions. With both ligands at 1 μM, which is nearly physiological in bacterial cells, close to half of the 30S PICs formed with IF2(GTP) and $\text{tRNA}^{\text{fMet}}$ arrive as a preformed ternary complex. With the high concentrations of both molecules in cells, preformation of a complex in solution may be an efficient mechanism that brings two large molecules onto the ribosome at once. Overall, this result suggests that multiple

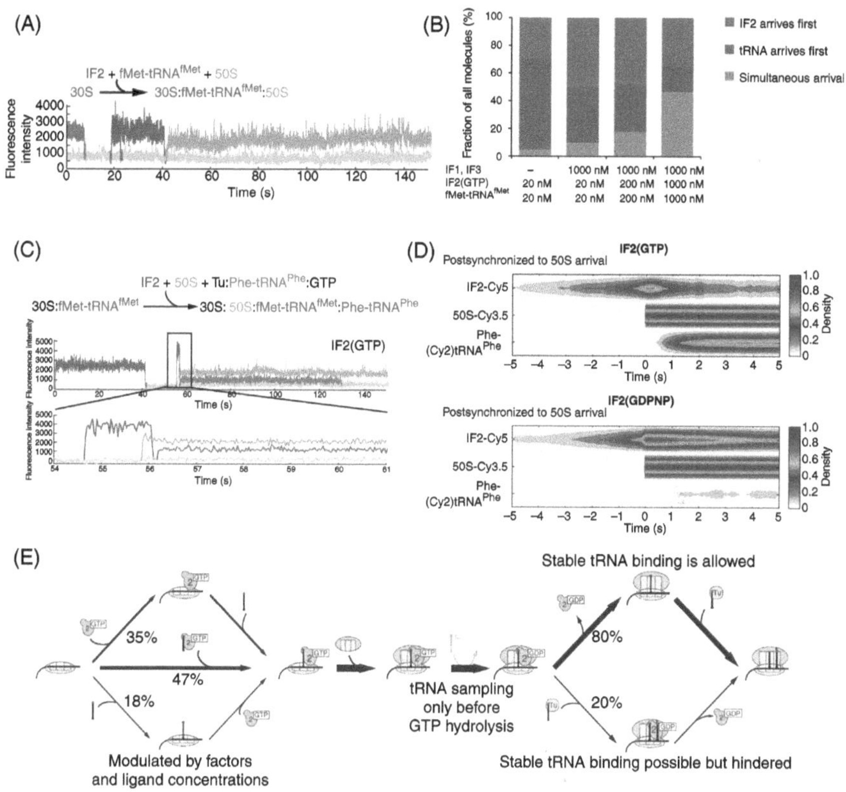

Figure 3 Tracking the assembly of a translationally competent ribosome complex through initiation. (A) With all of the essential components for prokaryotic translation initiation labeled, the entire initiation pathway from 30S PIC formation to subunit joining could be monitored in the ZMW. (B) All possible sequences of IF2 and initiator tRNA arrival was observed in the formation of functional 30S PICs. The preferred pathway leading to the assembly of a functional 30S PIC varied depending on the presence of other factors (eg, IF1 and IF3) and the concentrations of IF2 and initiator tRNA. At high concentrations of both, which is closest to the condition in vivo, simultaneous arrival of both molecules as a complex was the predominant mode of 30S PIC formation. (C) Tracking the ribosome's transition from initiation into elongation revealed the role of IF2 in controlling the arrival of the first elongator tRNA. (D) Postsynchronizing to the arrival of large 50S subunit (aligning all events to have 50S arrival to be $t = 0$) showed that IF2 departs within 2 s after subunit joining, followed by the arrival of the first elongator tRNA. When IF2 cannot hydrolyze GTP, as was the case with the nonhydrolyzable GTP analog GDPNP, it remained on the ribosome after subunit joining and blocked the arrival of the first elongator tRNA. Therefore IF2 departure subsequent to GTP hydrolysis is the final gating event for elongation to begin. (E) Investigating translation initiation at near-physiological conditions using the ZMW highlighted the complexity and the heterogeneous pathways that could potentially serve as points of regulation before the ribosome fully commits to translating a full protein. Nonessential factors, such as IF1 and IF3 likely modify the flux through each pathway, leading to further layers of regulation and fine tuning. *Adapted with permission from Tsai et al.*[83]

pathways are all valid for the formation of a functional PIC and that the flux through each is dependent on the concentration of the initiator tRNA and the IFs (Fig. 3E left panel). Subsequently, adding 50S subunits and the elongator tRNA in solution rapidly led to the formation of a 70S complex that then accepted the elongator tRNA in the presence of GTP (Fig. 3C). When the fluorescent signals from IF2-Cy5 and $tRNA(Cy2)^{Phe}$ were aligned (postsynchronized) to the arrival of 50S-Cy3.5 (Fig. 3D), there was a short overlap of about 2 s between the 50S and IF2 signal. The elongator tRNA also arrived about 2 s after subunit joining; this delay was independent of the concentration of the elongator tRNA, as long as tRNA binding does not require more than 2 s and becomes the rate-limiting step. Substituting GTP for a nonhydrolyzable analog, GDPNP, significantly lengthened the dwell time of IF2 on the ribosome postsubunit joining (Fig. 3D). Binding of the elongator tRNA became suppressed in this case. This line of evidence indicates that GTP hydrolysis or, more exactly, IF2 transitioning into its GDP-bound conformation, whose affinity to the ribosome is significantly lower[85] as seen in the short dwell time of the GDP-bound form of IF2, allows it to depart. Therefore IF2 departure followed by elongator tRNA arrival is the final gating event that marks the end of initiation and the beginning of elongation (Fig. 3E right panel); GTP hydrolysis is the preceding step necessary for dissociation by triggering conformational changes on the IF2[85] and the ribosome.[28]

11. THE ROLE OF EF-G IN TRANSLOCATING THE RIBOSOME: COUPLING COMPOSITIONAL DYNAMICS TO CONFORMATIONAL CHANGES OF THE RIBOSOME

Similar to initiation, translocation of the ribosome during elongation is also a process involving several stochastic steps in rapid succession and in coordination with major conformational changes. Translocation additionally involves major physical movements of the tRNA and mRNA. The two subunits of the ribosome can either adopt one of two global conformations when they are associated,[86,87] characterized by the relative rotation of the small subunit body to the large subunit[88]: (1) the nonrotated (posttranslocation) state where the ribosome is relatively devoid of local structural fluctuations and the tRNA with the nascent peptide chain in complex with the mRNA is firmly held in the P site and (2) the rotated (pretranslocation) state where the ribosome is unlocked to allow local conformational fluctuations, including letting the tRNAs in the A and P sites fluctuate partially into the next site of the 50S subunit. After peptide bond formation, when the ribosome is waiting for translocation to occur, it adopts the rotated conformation, loosening its grip on the tRNAs and the mRNA[27,89] in anticipation for the movements required to translocate precisely the A- and P-site tRNAs into the next site[60] and the mRNA by exactly 1 codon[61] with the tRNAs. The ribosome is intrinsically able to carry out all the required motions to translocate the tRNAs and mRNA, in addition to returning itself into the nonrotated state once translocation is completed, although this process requires about 3 h to occur spontaneously without assistance by factors.[90] EF-G, by taking advantage of the energy from GTP hydrolysis, greatly reduces the time of translocation to c. 50 ms at 1 μM of EF-G.[59,91] Previous studies suggest that EF-G binding in its GTP-bound form guides the conformational changes on the ribosome,[92,93] biasing ribosomal and tRNA conformations to favor translocation.[94–96] GTP hydrolysis is critical to the ability of EF-G to repeatedly drive translocation and the GDP-bound form of EF-G quickly dissociates from the ribosome posttranslocation. Single-molecule experiments[97–99] have additionally identified correlated conformational changes of the L1-stalk on the ribosome that serves as an opening and closing cap that only allows the E-site tRNA to depart after translocation. Nevertheless, the exact timing of translocation with respect to EF-G binding to the ribosome and GTP hydrolysis remained unclear. Depending on the actual sequence of events, the role of GTP hydrolysis may be directly linked to catalyzing translocation or to cycle the EF-G off from the ribosome, opening the A site for tRNA binding. Answering this question requires correlating signals reporting on both changes in ribosomal conformation and EF-G binding dynamics at a sufficiently high concentration of labeled EF-G. The experiment should also ideally be able to observe multiple rounds of translocation to characterize the effectiveness of EF-G binding in driving translocation.

Chen et al.[100] took advantage of the ZMW technology to track translocation with Cy5-labeled EF-G and a FRET signal that reports on the global rotational state of the ribosomal subunits[24,29,101] (Fig. 4A). The nonrotated ribosome exhibits a high-FRET state that changes into a lower FRET state when the two subunits rotate upon peptide bond formation. A subsequent counter rotation of the subunits during translocation returns the ribosome to the nonrotated or the high-FRET state. Thus each high–low–high FRET sequence represents the translation of a single codon on the mRNA. Although the original labeling scheme for the ribosomal subunits used

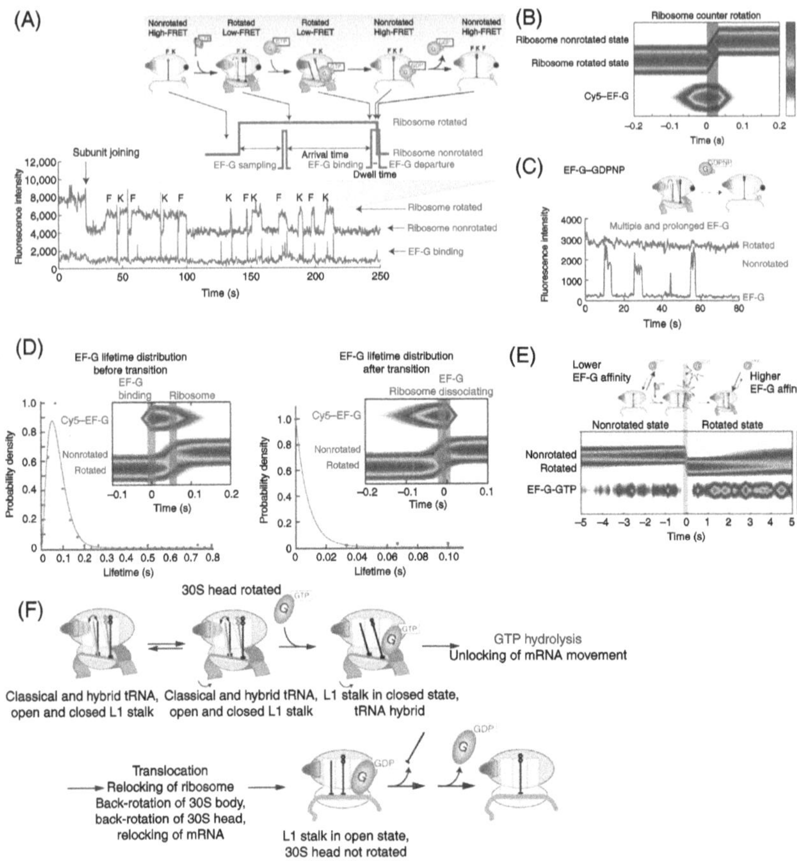

Figure 4 Dissecting the mechanism of EF-G catalyzed translocation. (A) With sufficiently high concentrations of Cy5-labeled EF-G [dark gray (red in the web version)], clear binding events could be seen coupled with intersubunit conformation change from the rotated state back to the nonrotated state, which is coupled to translocation. (B) Correlating the EF-G binding events to the exact time of intersubunit conformation change showed that EF-G binds approximately 70 ms prior to intersubunit counter rotation and dissociates rapidly afterward in about 30 ms. (C) When EF-G cannot hydrolyze GTP, as was the case with the nonhydrolyzable GDPNP, prolonged EF-G binding events that failed to drive translocation became frequent. (D) Fitting the dwell time of EF-G prior to intersubunit rotation/translocation yielded a Poisson distributed lifetime, indicating that there are multiple rate-limiting steps. GTP hydrolysis is likely one of them. Posttranslocation, the dwell time was exponential, suggesting only a single rate-limiting step is present, which is EF-G dissociation from the ribosome. (E) The affinity of the ribosome for EF-G changed depending on its rotational state. Upon transition from the nonrotated to the rotated state, EF-G sampling frequency increased as seen in the higher density in the EF-G-GTP row of the plot. (F) EF-G catalyzes translocation first by binding to the ribosome in the rotated state and then guiding the internal conformational fluctuations of the ribosome to the correct configuration for translocation. GTP hydrolysis then drives translocation and both the ribosome and EF-G undergo major conformational changes. Once translocation has completed, EF-G-GDP departs from the ribosome rapidly due to its low affinity for remaining bound to the ribosome. *Adapted with permission from Chen et al.*[100]

Cy3 and Cy5 as the FRET pair for the 30S and 50S subunit, respectively, ZMWs quenches Cy5 fluorescence nonlinearly[54] led to their substitution with BlackHole Quencher 2 (BHQ-2).[102] This change allows the tracking of the elongation cycle through intersubunit conformation changes using only a single color Cy3,[100,102] freeing the photophysically well-behaved Cy5 for other purposes. When ribosomes translated in the presence of Cy5-labeled EF-G with GTP, short pulses of EF-G binding could be seen mostly when the ribosome was in the rotated conformation (high Cy3 fluorescence or low-FRET state). An EF-G binding signal that coincided with the Cy3 signal changing from high to low intensity (low to high FRET) clearly correlated EF-G binding on the ribosome with translocation. Aligning the EF-G signal to the low-to-high FRET transition of the intersubunit conformation signal (Fig. 4B) showed that EF-G arrived about 60 ms prior to translocation. After the intersubunit conformation change, which lasted for 30 ms, EF-G quickly departed within 30 ms. Replacing GTP with a non-hydrolyzable analog, GDPNP (Fig. 4C), led to an increase of dwell time on the ribosome by more than 10-fold and a very low success rate for translocation. At physiological concentrations of free magnesium ions, which are about 1 mM, GTP hydrolysis increased the translocation rate by 50-fold. The GTP bound form of EF-G had significantly higher binding affinity to the ribosome in either conformational state, but the dissociation rate of EF-G from the ribosome only decreased when the ribosome was in the rotated state, selectively retaining the factor when the ribosome is in the right conformational state for translocation. A detailed examination of the lifetimes of EF-G on the ribosome (Fig. 4D) revealed that only EF-G bound immediately before translocation displayed a Poisson distribution indicative of approximately three rate-limiting steps within that process, with GTP hydrolysis likely being one of them. On the other hand, all other types of EF-G binding events, as well as the time needed for EF-G departure from the ribosome immediately after translocation, had lifetimes that are exponentially distributed; these lifetimes are determined by the single-step process of EF-G dissociating from the ribosome. These results suggest that EF-G hydrolysis, instead of just EF-G binding to the ribosome, is required to catalyze efficient translocation. The conformational state of the ribosome additionally regulates the binding affinity of EF-G (Fig. 4E) so that it preferentially binds in the pretranslocation state where it is needed. Interestingly, spectinomycin[29,99] and viomycin,[103,104] which are translocation-blocking antibiotics, changed EF-G binding lifetimes of events that did not lead to successful translocation on the rotated ribosome to Poisson distributed, suggesting that the translocation and the energy of GTP hydrolysis can be decoupled. This coordination between ribosome conformation and EF-G dynamics (Fig. 4F) is the key in maintaining the efficiency of translocation by ensuring that the energy of GTP hydrolysis is channeled at the right moment to catalyze the physical movement of the tRNAs, mRNA, and the ribosome subunits themselves.

Building upon Uemura et al.,[54] the two preceding works have fully taken advantage of the ability of the ZMW technology to work at high concentrations of fluorescent molecules in solution and to monitor simultaneously several compositional and conformational signals. Tsai et al.[83] enabled the authors to investigate the translation initiation pathway at near-physiological conditions through canonical pathways without the need for protocols to synchronize or freeze ribosomes into specific states along the way. Their observations highlight the heterogeneous nature of the initiation pathways available to assemble the 30S PIC and the crucial roles that the concentrations of different ligands and translation factors play in changing the flux through each of those possible options. The diverse pathways available during this stage could allow regulation to ensure that translation only occurs as required by the cell. The latter steps of initiation then focus on transitioning the ribosome from a stable complex over the start codon into a dynamically moving 70S complex along the rest of the mRNA. Direct observation of the process reveals that it involves the coordination of several molecular events (50S subunit binding, GTP hydrolysis on IF2, IF2 departure from the ribosome, and elongator tRNA arrival) within a short period of time, potentially providing a final checkpoint to ensure that the ribosome is ready to translate. Similarly, Chen et al.[100] observed that the elongating ribosome also relies on the tight coordination of several large-scale molecular transitions in order to translocate on the mRNA by precise 1-codon steps. To catalyze translocation using the energy of GTP hydrolysis, the ribosome must synergistically coordinate the binding of EF-G, GTP hydrolysis by EF-G, and the counter rotation of the ribosomal subunits. Depending on the intersubunit conformation, the ribosome preferentially increases the affinity of EF-G to the pretranslocation state with the subunits rotated after peptide bond formation and only triggers GTP hydrolysis when the local conformational fluctuations signal that the ribosome is ready to translocate. Any breakdown of these tightly linked processes, for example, when antibiotics are applied, disrupts translocation, leading to futile EF-G bindings and unnecessary GTP hydrolysis. These studies demonstrate that conducting single-molecule fluorescence experiments using the ZMW technology finally allows significant portions of the global dynamics of translation to be observed in their entirety. Moreover, the ribosomes observed can translate using canonical in vivo pathways, shedding light on how the individual steps might come together in near-physiological conditions and allowing experiments to move beyond model systems. These successes of using an experimental ZMW microscopy platform opened the door to further efforts to adapt a commercially available platform, making the technology potentially available on a wider scale.

12. ADAPTING A COMMERCIALLY AVAILABLE ZMW INSTRUMENT FOR GENERAL SINGLE-MOLECULE FLUORESCENCE EXPERIMENTS

The aforementioned studies employed custom-made experimental ZMW microscopes that are prototypes not available in the commercial market. The technical complexity of the ZMW technology means that the instrumentation of the microscopes[105–108] and the fabrication process[109] required to make ZMW chips with biologically compatible surfaces are not widely accessible. Limited access to a ZMW microscope restricts the number of experiments that can be done within a reasonable amount of time, which is undesirable when studying a biological system where the system must be probed under numerous conditions and configurations. Low availability and access also discourage more exploratory experiments needed to tackle complex systems with many unknowns. These restrictions of custom-built experimental ZMW platforms were roadblocks to the adoption of ZMW microscopy as a commonly used single-molecule fluorescence technique. Furthermore, these experimental setups can observe around 3000 ZMW wells at once. On average, only a 1000 of them would have an immobilized molecule to ensure that statistically only 1 molecule resides on the bottom of each occupied well.[54] Otherwise, multiple functioning molecules within a well complicate data analysis with convoluted signals and potential cross-interactions between the molecules. In experiments that require the cross-correlation of several different signals the number of molecules statistically exhibiting a particular behavior of interest could quickly drop to less than 100. These low numbers increase the statistical error in quantitation and impair attempts for more advanced and detailed analysis.

The commercial DNA sequencer (RS) platform developed by Pacific Biosciences (Menlo Park, California, USA) and the ZMW chips [single-molecule real-time (SMRT) cells] the instrument uses, on the other hand, are commercially available as a complete package. The instrument uses single-molecule fluorescence from dye-labeled nucleotides to sequence a piece of DNA when a polymerase immobilized in the ZMW well replicates it; the RS platform is essentially an automated single-molecule microscope with the necessary fluidics to conduct dynamic experiments involving real-time delivery of components. Coupled with the high throughput of its ability to observe from 75,000 to 150,000 ZMWs simultaneously, repurposing the instrument to be a general purpose single-molecule fluorescence microscope would significantly improve the availability of the ZMW technology beyond next-generation sequencing (Fig. 5A).

13. THE RS SEQUENCER PROVIDES A FLEXIBLE PLATFORM FOR MULTICOLOR AND HIGH-THROUGHPUT SINGLE-MOLECULE MICROSCOPY

Chen et al.,[110] in collaboration with members of the engineering team of Pacific Biosciences, developed a software platform and a workflow that could potentially work with any RS sequencer with only software updates to the instrument. The hardware setup of the instrument includes two lasers at 532 nm (green) and 642 nm (red) and spectral channels that roughly corresponds to the emission spectrum of commercially available cyanine dyes Cy3, Cy3.5, Cy5, and Cy5.5, all of which possess good photophysical properties and long photobleaching lifetimes. The instrument has an automated alignment procedure and a six-axis stabilized stage that permits stable imaging over hours. A constant flow of nitrogen and a layer of mineral oil prevent evaporation of the liquid reagents within a ZMW chip. They also exclude atmospheric oxygen from the ZMW chip, reducing the possibility of premature photobleaching of the dyes through reaction with oxygen radicals. The nitrogen atmosphere in the imaging chamber also increases the effectiveness of any oxygen scavenging system because it no longer has to deal with an additional influx of atmospheric oxygen.

When the RS functions as a sequencer, the automated workflow performs most of the necessary tasks to prepare and load the sample onto the ZMW chip; however, the workflow involves a hard-coded 1.5-h preparation time unsuitable for labile reagents and is inflexible toward customization that many experiments may require. The optical alignment process lasts another

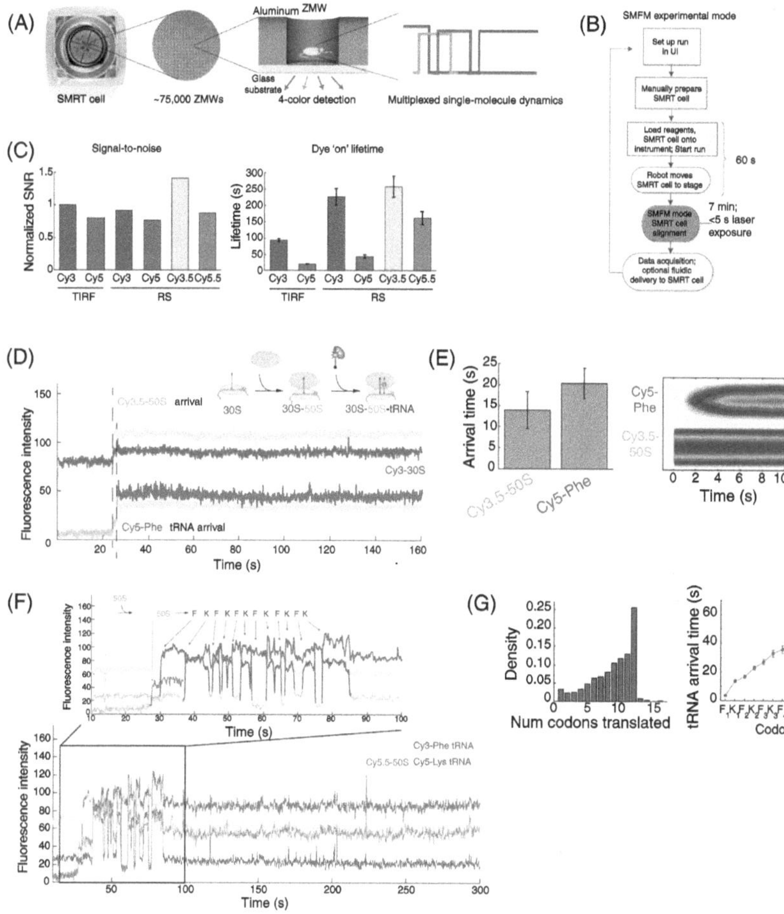

Figure 5 A modified RS platform for use as a general single-molecule fluorescence microscope. (A) The commercial RS instrument uses the SMRT cell containing 75,000–150,000 ZMW wells, all of which can be imaged simultaneously in four different color channels. (B) The modified workflow for single-molecule fluorescence imaging is flexible toward different experimental protocols involving live deliveries of solutions into the chip during imaging and minimizes unduly exposing the ZMW chip to laser illumination prior to the experiment. (C) The four color channels, corresponding to the commercially Cy3, Cy3.5, Cy5, and Cy5.5 dyes exhibited comparable signal-to-noise ratios compared to a conventional TIRF microscope. The oxygen-free environment inside the imaging chamber significantly improved the photobleaching lifetimes of the dyes compared to an unsealed microscope. (D) Repeating the initiation experiments of Tsai et al.[83] on the RS yielded comparable results. (E) The times required for subunit joining and elongator tRNA arrival compared favorably with the original experiments. (F) Repeating the tRNA transit experiment of Uemura et al.[54] with more than 3000 observed molecule was possible with the high throughput of the RS platform. The four-color capability additionally allowed both ribosomal subunits to be tracked. (G) The high concentration of labeled tRNAs used ensure the most molecules translated the full 12-codon mRNA, with fast translation rates for single-molecule experiments on the order of a few seconds per codon. *Adapted with permission from Chen et al.*[110]

15 min with laser light exposure of about 2 min, which can photobleach a significant portion of even photostable dyes; most wells optimally loaded with a single dye-labeled molecule would be dark by the time the experiment starts. Chen et al. created a new customizable single-molecule workflow (Fig. 5B) that requires only 60 s of instrument preparation time and minimal laser exposure during the subsequent alignment step. The new software interface allows for key experimental parameters, including acquisition time, laser power, camera exposure time, and fluidics parameters to be customized using a graphical user interface. These changes were all done through software modifications to the instrument, without any hardware changes. The minimal levels of modification allows any RS sequencer with the proper software to be switched quickly between single-molecule microscopy mode and sequencing mode—the same instrument can do double duty as both a general single-molecule microscope and a sequencer.

Using the modified workflow, Chen et al. characterized the basic performance of the RS instrument using the four commercial cyanine dyes mentioned earlier (Cy3, Cy3.5, Cy5, and Cy5.5; Fig. 5C) to ensure that their photophysics is comparable to that of a conventional TIRF microscope. With the dyes attached to a biotinylated DNA oligo and the oligo immobilized in the ZMW via neutravidin, the signal-to-noise ratio of all dyes were good and directly comparable to that of a standard TIRF setup. Using the same PCA/PCD oxygen scavenging system[111] to remove dissolved oxygen in solution in both the RS and the TIRF microscope, the nitrogen atmosphere in the RS instrument further improved the lifetimes before photobleaching by a factor of two. At the laser power required to image at 10 frames per second, Cy3, Cy3.5, and Cy5.5 on average lasted 3–4 min. Given the large number of ZMW wells per chip (150,000), the instrument could potentially track a statistically significant number of dye-labeled molecules bound in the ZMW for up to 10 min. Using a slower frame rate and a correspondingly lower laser power, even longer timescales of close to an hour are within reach. This time window places a large number of complex biological processes within reach for observation. After characterizing the basic performance of the RS, Chen et al. repeated previous experiments to verify that the instrument is capable of performing complex and biologically relevant single-molecule experiments. The authors observed elongation using the intersubunit FRET signal[100,102] and the arrival of the first elongator tRNA after 50S subunit joining to a 30S PIC as conducted by Tsai et al.[83] (Fig. 5D); the results agreed with the prior data (Fig. 5E). The improved photophysics of the instrument significantly increased the number of ribosomes translating the full 12 codons on the mRNA for the elongation experiment and the lifetimes of all of the labeled ligands in the initiation experiments were also longer than the original work. A later work by Capece et al.[112] was successful in synthesizing functional SNAP tags from translating ribosomes in ZMWs, demonstrating that complex biological processes can take place within the confines of ZMW wells. Finally, to showcase the potential throughput of the instrument, Chen et al. repeated a version of the tRNA transit experiment of Uemura et al.[54] but with more than 3300 molecules analyzed (Fig. 5F and G), which is 10-fold more than the original experiment. With the significantly higher number of ZMW wells on a chip, the limiting factor to the practical output of single-molecule experiments on the RS would shift from the time required to run experiments in order to capture a statistically sufficient number of molecules onto the speed of the analysis pipeline; automating previously manual analysis processes would be the key to utilizing the full throughput of the RS platform. This work demonstrates that the RS is a fully capable single-molecule fluorescence microscope, whose advanced instrumentation yields a significant improvement to dye behavior. Dissecting the mechanism of complex biological processes that require extended observation over many stochastic steps and ones that involve heterogeneous pathways would take full advantage of the capabilities the RS offers.

14. USING THE RS TO DISSECT THE MECHANISM OF TRANSLATIONAL STALLING

Two ensuing studies on how realistic and complicated mRNA sequences influence ribosomal dynamics further highlighted advantages of using the RS as a single-molecule fluorescence microscopy platform. First, Tsai et al.[113] dissected the mechanisms of translational stalling that involves complex interactions between the nascent peptide and the exit tunnel of the large 50S subunit. The SecM stall sequence is a part of the bacterial secretory pathway that monitors the secretion capacity of the cell: a sufficient number of translocons relieves stalling in a timely manner whereas extended stalling by SecM upregulates the expression of the translocase SecA that is cotranscribed on the same mRNA.[114,115] Bulk biochemical studies have identified the 17-amino acid minimal stall sequence that is completely contained within the peptide exit tunnel of the 50S subunit when stalling occurs[116]; stalling of translation depends solely on peptide–ribosome interactions without the need for cofactors.[117,118] This challenges the previous assumption that the exit tunnel is an inert passage way that allows nascent peptides to pass through with minimal interactions.[119] The arginine at position 15 and the proline at 18 are specifically required, indicating that they are the key residues that interact with the rRNA and ribosomal proteins lining the exit tunnel.[116] The amino acids further upstream compacts the nascent chain in the tunnel, likely setting the stage for Arg and Pro to interact with the exit tunnel.[120] A cryoelectron microscopy structural study has captured a low-resolution structure of stalled ribosomes, whose P site contains a Pro tRNA and whose peptidyl transferase center (PTC) has been remodeled to disrupt catalyzing peptide bond formation.[121] Molecular dynamic simulations have proposed a possible mechanism that involves specific networks of interactions between the nascent peptide and the exit tunnel.[122] However, without observing the behavior of the ribosome as it is translating the SecM sequence, the mechanistic picture for the dynamic pathways of stalling remained incomplete. Drawing conclusions on the mechanism of stalling predominantly from the final stalled ribosome provides no information on how a dynamically translating ribosome becomes a stalled one.

Using a translation system containing a full complement of tRNAs aminoacylated using tRNA synthetases purified from *Escherichia coli* extracts,[123] Tsai et al. observed translation of the entire minimal SecM stall sequence using the intersubunit FRET signal (Fig. 6A). While most ribosomes stalled near the final proline in the stall sequence, a significant number translated up to 4 codons beyond the end of the stall sequence before stalling completely. A labeled tRNA for a Phe codon that is 3 codons downstream of the final Pro18 in the stall sequence showed an extensively lengthened dwell time, further supporting that ribosomes are stalling past Pro18. These results indicate that instead of a single stall site, translation slows down significantly over a stretch of 4 codons. A small fraction of ribosomes eventually cleared the stall sequence without external help, resuming normal rates of translation of 4–5 codons after the end of the stall sequence. Tsai et al. estimated that this placed the half-life of stalling to be about an hour, in line with estimations using data from previous studies.[124,125] Examining the lifetimes of the two conformational states of the ribosome on the SecM mRNA (Fig. 6B) revealed that elongation rates began to slow down 2 codons after the critical Arg15 is translated. Mutating the minimal stall sequence showed that Pro18 is needed to precipitate the final stalling (Fig. 6C) and Arg15 induces the initial slowdown in the nonrotated state (Fig. 6D). Replacing Arg15 with similarly charged residues, shifting the location of Pro18 by 1 codon, and replacing

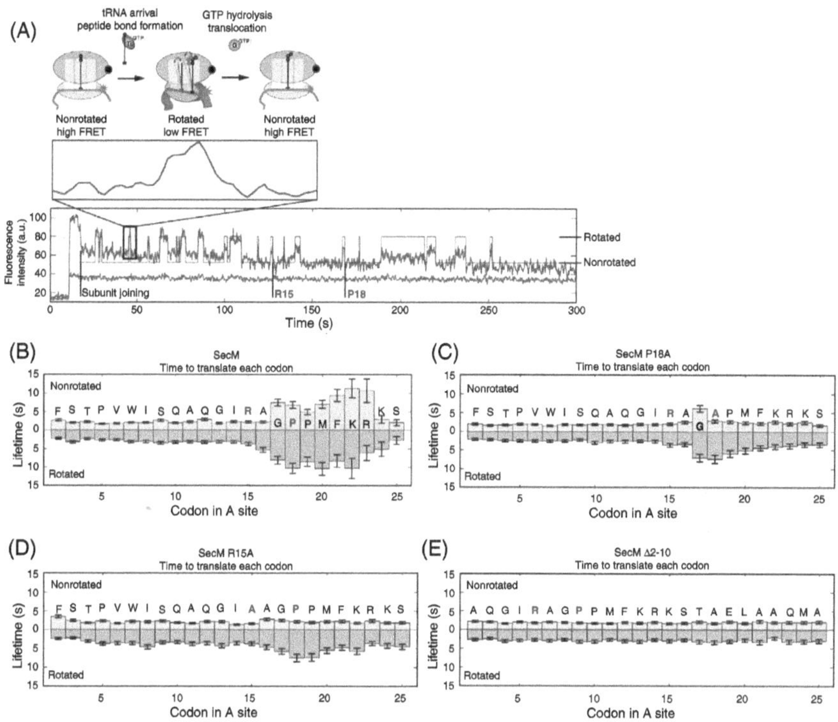

Figure 6 Probing interactions between the nascent peptide and the exit tunnel critical in translational stalling. (A) Tracking the intersubunit conformation as the ribosome is translating an mRNA coding for the SecM stall sequence revealed that the stall sequence induces slowdowns over specific locations on the mRNA. (B) The wildtype SecM sequence induced a general slowdown in translocation as shown in the longer rotated state lifetimes after the first critical Arg15 has entered the exit tunnel. The nonrotated lifetimes also increased over codon 17 and 2 codons after the final Pro18, suggesting difficulties in peptide bond formation. (C) Mutating Pro18 to Ala removed all slowdowns in the nonrotated state lifetimes after the 18th codon and no stalling ensued. (D) Changing Arg15 to Ala removed all inhibitions in the rate of peptide bond formation while some decrease in the translocation rate could still be seen. No stalling was observed. (E) Removing the first nine amino acids of the stall sequence, while retaining the critical Arg and Pro, returned translation dynamics completely back to that comparable with nonstalling mRNA sequences. (F) The specific effects of each element on the minimal stall sequence on ribosome dynamics show that stalling is a dynamic event involving several precisely timed events in a cascade. The amino acids prior to the critical Arg15 induces peptide compaction in the tunnel to reduce translocation rate, allowing Arg15 to interact with the tunnel. Arg15 then remodels the PTC via a network of interactions starting from within the exit tunnel, slowing peptide bond formation. Pro18 then precipitates stalling through its unique structural property, locking the nascent peptide in conformations that severely reduces peptidyl transfer rates and translocation rates. *Adapted with permission from Tsai et al.*[113]

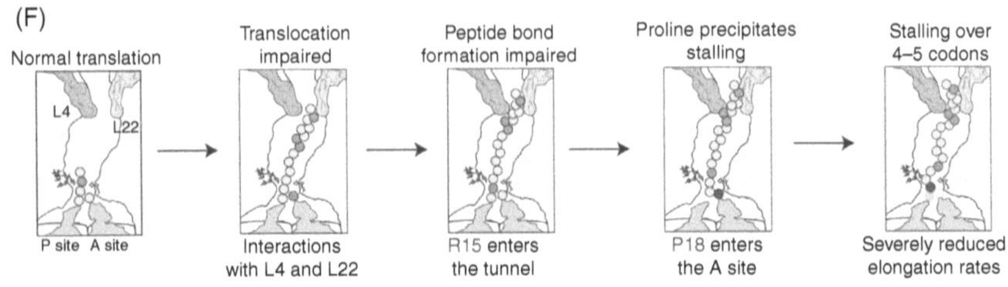

Figure 6 *(Continued)*

proline with a structurally similar analog azetidine all failed to rescue stalling, demonstrating that both amino acids are specifically required at exactly the right place. The upstream sequence is also important in setting the stage for Arg15 and Pro18 to interact with the exit tunnel (Fig. 6E) by slowing down translocation through peptide compaction; its removal abolished stalling altogether as well as any signs of translational slowdown. These results paint a dynamic picture of stalling (Fig. 6F), where one precisely positioned interaction between the exit tunnel and the nascent peptide sets the stage for the next. This cascade of events work cooperatively to place the critical amino acids in the right location with respect to specific residues of the rRNA on the large subunit. A network of interacting rRNA residues transduce this mechanical stress to change to the shape of the PTC, disrupting peptidyl transfer on the ribosome and stalling translation. Moreover, SecM-induced stalling is better envisioned as a significant slowdown in dynamics lasting over a defined amount of time rather than ribosomes frozen in a particular conformational state; sufficient external or internal force that breaks these interactions allows translation to resume, which is the proposed mechanism of how the translocon relieves stalling.[126]

Johansson et al.[127] investigated another class bacterial stall sequence, *ermCL*, which similarly precipitates stalling through interactions within the exit tunnel. While the peptide sequence of SecM alone is sufficient to cause stalling, ErmCL additionally requires antibiotics of the macrolide class as cofactors. Ribosomes stalled in the presence of macrolides then open a hidden ribosome binding side downstream, leading to the translation of the *ermC* methyltransferase that confers antibiotic resistance.[128] Traditionally, macrolides were thought to be a physical plug that prevents the ribosome from extruding the nascent peptide, eventually leading to the peptidyl-tRNA dissociating from the ribosome.[129] However, recent studies have shown that these antibiotics bind inside the exit tunnel near the PTC, contacting both rRNA residues and ribosomal proteins.[130] The action of ErmCL, similar to

SecM, requires a specific peptide sequence, with the sixth to the ninth amino acid being critical. These results suggest a stall mechanism that involves well-defined interactions between the nascent peptide, ribosome and macrolide.[131] Peptides that elude the effects of macrolides or evict the drugs altogether from the exit tunnel further challenged the idea of an inert physical plug.[132,133] While both SecM and ErmCL require specific interactions between the nascent peptide and the exit tunnel to stall the ribosome, Johansson et al. observed that ErmCL stalls the ribosome abruptly with the 10th codon in the P site of the ribosome. Using the intersubunit FRET signal, the authors showed that ErmCL did not affect translocation rates in the presence of erythromycin but increased the time required for peptide bond formation by fourfold only over the stall site. When observed using the RS instrument, ErmCL-stalled ribosomes displayed multiple unproductive tRNA samplings to the A site when stalled, suggesting that the ribosomes cannot transfer the nascent peptide chain from the P-site tRNA onto the A-site tRNA because of a defective PTC conformation. The *h-ns* sequence that is able to bypass the blocking effects of erythromycin showed yet another type of dynamic slow-down: its effect was confined to slowing the translocation rate over 1 codon. Interestingly, similar to SecM, more EF-G binding attempts were needed to resolve the rotated state, implying an increased energy barrier to translocation. By channeling the ribosome into the rotated state, characterized by numerous spontaneous local conformational fluctuations, *h-ns* is able to bypass the effects of macrolides and continue to translate. Thus investigating different sequences that involve specific interactions in the exit tunnel led to significantly different dynamic behaviors, suggesting fundamental differences in their mechanisms that exploit different aspect of the elongation cycle. SecM leads to a gradual decrease in both peptide bond formation and translocation rates over several codons with ribosomes stalling in a collection of different states. Translation slowdown in ErmCL and *h-ns* is abrupt over a specific codon, where ErmCL specifically disrupts peptide bond formation and *h-ns* increases the energy barrier to translocation.

15. THE MECHANISM OF −1 FRAMESHIFTING

While Tsai et al. and Johansson et al. examined stalling through nascent peptide interactions with the exit tunnel, Chen et al.[134] employed the RS to understand the mechanism of −1 frameshifting where the ribosome must physically shift back one nucleotide on the mRNA into a different reading frame. While spontaneous frameshifting in translation is normally rare (10^{-3}–10^{-4} per codon)[135] and usually detrimental, specific slippery mRNA sequences (X XXY YYZ, where the same letter denotes the same nucleotide) in conjunction with mRNA secondary structures can induce significantly elevated levels of programmed frameshifting by one nucleotide backward (toward the 5′ direction of the mRNA).[136,137] A 3′-harpin[138] and a 5′-internal-Shine–Dalgarno sequence (in some prokaryotes)[139] that pause the ribosome over the slippery sequence are additionally needed but insufficient by themselves to induce frameshifting. Taking advantage of the potential ambiguity that the slippery sequence provides in the −1 reading frame, where only the wobble position changes compared to the 0 reading frame, the anticodons of the tRNAs must detach from the mRNA and reassociate in the new −1 reading frame. Despite extensive work to dissect frameshifting, the exact timing and therefore the position of the ribosome on the mRNA during slippage remained elusive.[140,141] As a comprehensive mechanism was not available, the factors that control the fraction of ribosome that frameshift, which could vary from 1% to 80%,[142] also was unclear. Different proposed mechanisms placed the exact timing of the frameshift at different steps along the elongation pathway: (1) during accommodation of the A-site tRNA,[143] (2) postaccommodation but prior to peptide bond formation,[140] (3) during EF-G catalyzed translocation,[142,144] or (4) post-translocation.[141] Direct observation of the dynamics of frameshifting would provide the key pieces of information that could distinguish between the four different proposed models; however, the need to track the conformational state of the ribosome simultaneously with tRNA and EF-G binding has also made investigating the dynamics of frameshifting through single-molecule methods difficult.

Chen et al. observed the translational dynamics on the *dnaX* frameshifting sequence (the specific slippery sequence is -A AAA AAG-) in real time. The mRNA sequence for this gene contains the slippery sequence -A AAA AAG-, as well as an internal Shine–Dalgarno sequence and a hairpin flanking the slippery sequence (Fig. 7A). Taking advantage of the capabilities of the RS platform, the authors tracked the frameshifting process from multiple perspectives: intersubunit rotation, tRNA binding over and near the slippery sequence, and EF-G binding. By using an mRNA whose open reading frame is either 12 codons long without frameshifting or 9 codons with frameshifting, the observed frameshifting rate was approximately 75% (as tracked by intersubunit rotation and tRNA binding over the final codon), consistent with previous studies[138,139] (Fig. 7B). Upon closer inspection, the ribosomes translated in two distinct profiles: ribosomes that showed short rotated state lifetimes and translocated quickly over the slippery sequence starting from the 7th codon (Lys) did not frameshift at all, whereas ribosomes that slowed down over the same site subsequently frameshifted with a high probability. Disrupting the downstream hairpin or the internal Shine–Dalgarno sequence reduced the fraction of frameshifted ribosomes, as well as the severity of the pause over the slippery sequence for those that still underwent frameshifting. Both features likely induce the pause by increasing the energy barrier to translocation: the internal Shine–Dalgarno sequence by base pairing with the anti-Shine–Dalgarno sequence on the 16S rRNA and the hairpin by forcing the ribosome to physically open it. Thus while the pause over Lys7 is a hallmark of frameshifting, the first decision point occurs earlier than previously expected, prior to the initial pause in translation.

The translocation time at Lys7 codon as measured using the departure of a labeled tRNA immediately upstream of the slippery sequence revealed that it is significantly shorter than the rotated state lifetimes of frameshifting ribosomes. Furthermore, long tRNA sampling attempts appeared while the paused ribosomes were in the rotated state (Fig. 7C), a clear signal that the ribosome has translocated and the A site is open for tRNA binding. The time needed before tRNAs begin to sample the A site after translocation (measured from the time of E-site tRNA departure) was dependent on EF-G concentration (Fig. 7D). These pieces of evidence all point to the ribosomes pausing in a noncanonical conformation where the subunits failed to counter rotate after translocation.

Within this noncanonical posttranslocation state, changing the nucleotides of the slippery sequence in the A site of the ribosome altered the partition between ribosomes that do not stall, ones that stall but do not frameshift, and ones that stall and frameshift. Mutating -A AAA AAG- to -A AAA AAA- strengthens the wobble base interaction in the 0 reading frame to $tRNA^{Lys}$ (with anticodon UUU) compared to the −1 frame, leading to only half of the stalled ribosomes shifting frames. Changing the sequence to A AAA UUU, which eliminates the possibility of slippage, further decreased the percentage of frameshifting ribosomes to 20%. Given these results, the mechanism of frameshifting likely involves A-site tRNA-induced slippage in the noncanonical posttranslocation state. The strained ribosomal conformation with a looser grip on the tRNAs and the preference of $tRNA^{Lys}$ to base pair with AAA in the −1 frame instead of AAG in the 0 frame then allow the prolonged A-site tRNA sampling events

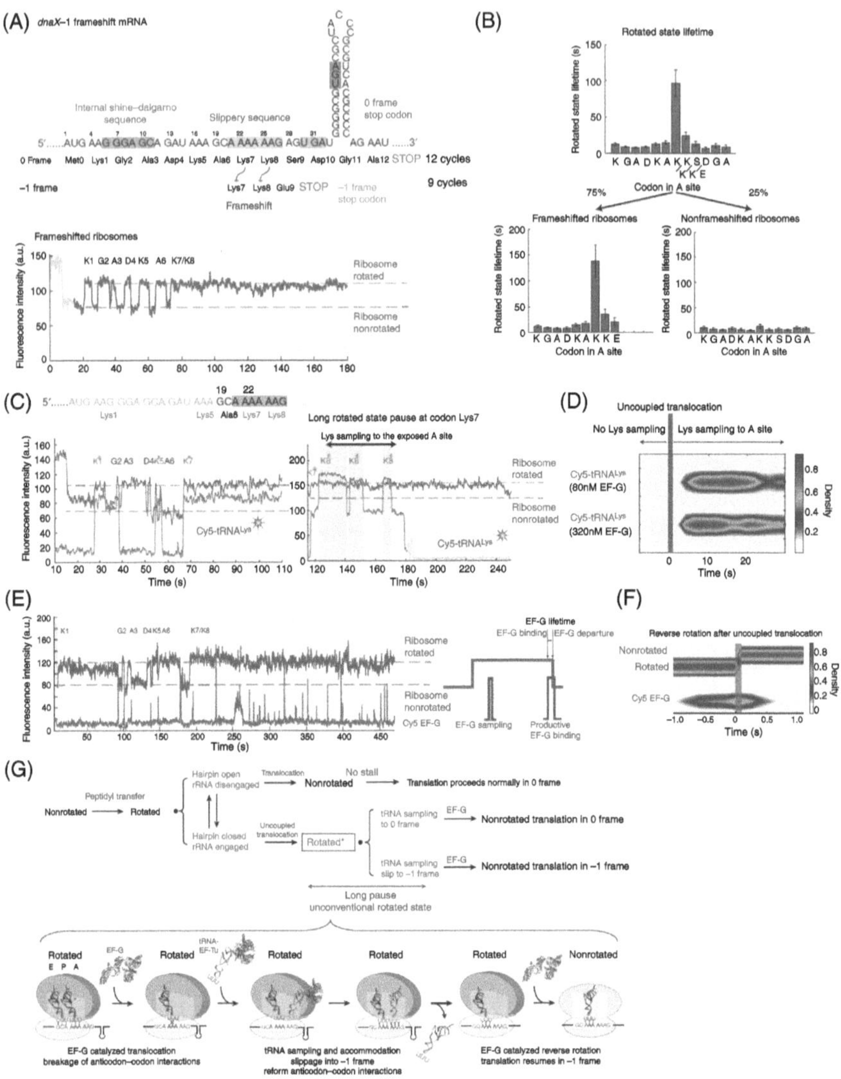

Figure 7 The mechanism of −1 frameshifting involves coordination between tRNA, EF-G, and intersubunit conformational changes. (A) The *dnaX* sequence contains all canonical elements present in sequences that induce −1 frameshifting: an optional internal Shine–Dalgarno sequence, an ambiguous slippery sequence (X XXY YYZ) and a hairpin afterward. The specific sequence used has an open reading frame of 9 codons for frameshifted ribosomes and 12 codons for nonframeshifted ribosomes, so the fraction of frameshifted ribosomes could be separated based on how many codons they translated. (B) The native *dnaX* sequence induced a slowdown in the rotated state specifically over codon 7 over the slippery sequence. While this pause was the hallmark of −1 frameshifting occurring for 75% of the time, ribosomes that did not frameshift showed no signs of pausing. (C) Monitoring tRNA binding dynamics over the

to redefine the reading frame. This A-site sampling serves as the second and final branch point that decides if frameshifting occurs. Chen et al. showed with Cy5-labeled EF-G that the noncanonical state resists EF-G from completing the counter rotation to return the subunits to their normal post-translocational state, as evidenced by the repeated sampling attempts from EF-G (Fig. 7E). Eventually, the elongation factor resolves the noncanonical state, returning the ribosomes to normal translation (Fig. 7F). Therefore while frameshifting itself takes place after translocation, a series of coordinated events starting from the initial pausing of translation are necessary to setup the noncanonical state so that the stage is set for the slippage of the tRNA–mRNA interaction to occur (Fig. 7G).

16. THE FUTURE OF ZMW MICROSCOPY IN THE STUDY OF COMPLEX BIOLOGICAL SYSTEMS

The works described earlier highlight the need to track directly the time evolution or dynamics, of the ribosome with its ligands to dissect the mechanism of translation. For initiation, the tight coordination of translation factor association and dissociation, biochemical reactions, subunit joining, and tRNA arrival ensures that a functional ribosome complex is assembled and primed to translate. During elongation, tRNAs transit through the

◀ slippery sequence revealed prolonged tRNA binding to the A site, perhaps after translocation with the ribosome in a noncanonical rotated state. These sampling events likely redefine the reading frame over the slippery sequence, allowing the ribosome to frameshift because the noncanonical state has a loosened grip on the tRNAs and the mRNA. (D) The time until the appearance of tRNA sampling was directly related to the concentration of EF-G, implying that while the ribosome does not return to its canonical nonrotated state, tRNA can sample posttranslocation. (E) EF-G was critical in resolving the noncanonical rotated state, where multiple EF-G sampling events usually preceded the final binding event that successfully counter rotated the subunit. (F) The final binding event counter rotated the subunit, but the EF-G resided for a prolonged period of time compared to normal translocation, indicating that a significant energy barrier is present. (G) The mechanism of −1 frameshifting involves several branch points along the way: first the ribosome engages the 3'-hairpin and optionally the 5'-internal Shine–Dalgarno sequence in prokaryotes. An uncoupled translocation then leaves the ribosome in a noncanonical rotated state, allowing tRNAs to sample in the A site. Prolonged A-site sampling over the slippery sequence leads to a redefinition of the reading frame, inducing a −1 frameshift. Finally, an EF-G resolves the noncanonical state, returning the ribosome to normal translation. *Adapted with permission from Chen et al.*[134]

ribosome in a coordinated manner, which is controlled by peptide bond formation and translocation. Both of those events are tightly coupled to major conformational rearrangements of the ribosome. EF-G catalyzes translocation by guiding these rearrangements and then accelerating translocation through the energy of GTP hydrolysis. Alternative translation processes such as stalling and frameshifting must modify ribosomal dynamics to accomplish their specific goals. For both stalling and frameshifting, a series of well-choreographed events must take place in the correct temporal sequence, setting the stage for the final dramatic event to take place; observations focused on the end state would miss the dynamic nature of stalling and frameshifting. Dissecting the aforementioned systems is challenging for conventional bulk techniques due to the presence of multiple dynamic and stochastic steps, in addition to the potential heterogeneity of the system.

The global dynamics of many biological processes that require minutes to occur is a significant challenge to single-molecule fluorescence methods that are limited to low concentrations of free-labeled components. As the works reviewed in this chapter showcase, the high concentrations of freely diffusing labeled components that are tolerated in ZMW microscopy opens the door to detailed investigation of realistic biological systems. The ease of use and the multiplexing capability of the RS platform additionally allowed for complex processes such as translational stalling and frameshifting to be observed in their entirety, making new discoveries on their mechanisms in the process. As the modified RS platform allows custom experimental protocol modifications, researchers could individually adapt the instrument to suit the requirements of most single-molecule fluorescence experiments. The ZMW technology has the potential to serve as a bridge between simplified model systems and difficult-to-control in vivo studies: it allows the investigation of complex processes at near-physiological conditions while still maintaining a controlled in vitro environment. Understanding how biological systems evolve over timescales on the order of minutes frames the overall process, disentangling its complexity. By combining detailed structural and kinetic information synergistically, a coherent mechanism can emerge. Moreover, the ability to track complex pathways in their entirety directly in real time could reveal information that reconciles conflicting hypotheses stemming from piecemeal observations. Observing biological processes close to their native contexts would greatly improve our understanding of how they actually work within living cells and organisms. ZMW technologies thus have a powerful role in the future of biophysics.

REFERENCES

1. Maaløe O, Regulation of the protein-synthesizing machinery—ribosomes, tRNA, factors, and so on. Goldberger R, ed. *Biological Regulation and Development*. vol. 1. US: Springer; 1979:487–542.
2. Warner JR. The economics of ribosome biosynthesis in yeast. *Trends Biochem Sci*. 1999;24(11):437–440.
3. Schaechter M, Maaloe O, Kjeldgaard NO. Dependency on medium and temperature of cell size and chemical composition during balanced grown of *Salmonella typhimurium*. *J Gen Microbiol*. 1958;19(3):592–606.
4. Kjeldgaard NO, Maaloe O, Schaechter M. The transition between different physiological states during balanced growth of *Salmonella typhimurium*. *J Gen Microbiol*. 1958;19 (3):607–616.
5. Watson JD. The synthesis of proteins upon ribosomes. *Bull Soc Chim Biol*. 1964;46: 1399–1425.
6. Marintchev A, Wagner G. Translation initiation: structures, mechanisms and evolution. *Q Rev Biophys*. 2004;37(3–4):197–284.
7. Agirrezabala X, Frank J. Elongation in translation as a dynamic interaction among the ribosome, tRNA, and elongation factors EF-G and EF-Tu. *Q Rev Biophys*. 2009;42 (3):159–200.
8. Petry S, Weixlbaumer A, Ramakrishnan V. The termination of translation. *Curr Opin Struct Biol*. 2008;18(1):70–77.
9. Kapp LD, Lorsch JR. The molecular mechanics of eukaryotic translation. *Annu Rev Biochem*. 2004;73:657–704.
10. Green R, Samaha RR, Noller HF. Mutations at nucleotides G2251 and U2585 of 23 S rRNA perturb the peptidyl transferase center of the ribosome. *J Mol Biol*. 1997;266 (1):40–50.
11. Johansson M, Bouakaz E, Lovmar M, Ehrenberg M. The kinetics of ribosomal peptidyl transfer revisited. *Mol Cell*. 2008;30(5):589–598.
12. Johansson M, Ieong KW, Trobro S, et al. pH-sensitivity of the ribosomal peptidyl transfer reaction dependent on the identity of the A-site aminoacyl-tRNA. *Proc Natl Acad Sci USA*. 2011;108(1):79–84.
13. Johansson M, Zhang J, Ehrenberg M. Genetic code translation displays a linear trade-off between efficiency and accuracy of tRNA selection. *Proc Natl Acad Sci USA*. 2012;109 (1):131–136.
14. Noller HF. Structure of ribosomal RNA. *Annu Rev Biochem*. 1984;53:119–162.
15. Clemons Jr WM, May JL, Wimberly BT, McCutcheon JP, Capel MS, Ramakrishnan V. Structure of a bacterial 30S ribosomal subunit at 5.5 A resolution. *Nature*. 1999;400 (6747):833–840.
16. Ban N, Nissen P, Hansen J, Moore PB, Steitz TA. The complete atomic structure of the large ribosomal subunit at 2.4 A resolution. *Science*. 2000;289(5481):905–920.
17. Ben-Shem A, Garreau de Loubresse N, Melnikov S, Jenner L, Yusupova G, Yusupov M. The structure of the eukaryotic ribosome at 3.0 A resolution. *Science*. 2011;334 (6062):1524–1529.
18. Selmer M, Dunham CM, Murphy FVJV, et al. Structure of the 70S ribosome complexed with mRNA and tRNA. *Science*. 2006;313(5795):1935–1942.
19. Voorhees RM, Ramakrishnan V. Structural basis of the translational elongation cycle. *Annu Rev Biochem*. 2013;82:203–236.
20. Julian P, Milon P, Agirrezabala X, et al. The Cryo-EM structure of a complete 30S translation initiation complex from *Escherichia coli*. *PLoS Biol*. 2011;9(7):e1001095.
21. Voorhees RM, Schmeing TM, Kelley AC, Ramakrishnan V. The mechanism for activation of GTP hydrolysis on the ribosome. *Science*. 2010;330(6005): 835–838.

22. Aitken CE, Petrov A, Puglisi JD. Single ribosome dynamics and the mechanism of translation. *Annu Rev Biophys.* 2010;39:491–513.
23. Petrov A, Chen J, O'Leary S, Tsai A, Puglisi JD. Single-molecule analysis of translational dynamics. *Cold Spring Harb Perspect Biol.* 2012;4(9):a011551.
24. Dorywalska M, Blanchard SC, Gonzalez RL, Kim HD, Chu S, Puglisi JD. Site-specific labeling of the ribosome for single-molecule spectroscopy. *Nucleic Acids Res.* 2005;33 (1):182–189.
25. Cornish PV, Ermolenko DN, Noller HF, Ha T. Spontaneous intersubunit rotation in single ribosomes. *Mol Cell.* 2008;30(5):578–588.
26. Petrov A, Puglisi JD. Site-specific labeling of *Saccharomyces cerevisiae* ribosomes for single-molecule manipulations. *Nucleic Acids Res.* 2010;38(13):e143.
27. Blanchard SC, Kim HD, Gonzalez Jr RL, Puglisi JD, Chu S. tRNA dynamics on the ribosome during translation. *Proc Natl Acad Sci USA.* 2004;101(35):12893–12898.
28. Marshall RA, Aitken CE, Puglisi JD. GTP hydrolysis by IF2 guides progression of the ribosome into elongation. *Mol Cell.* 2009;35(1):37–47.
29. Aitken CE, Puglisi JD. Following the intersubunit conformation of the ribosome during translation in real time. *Nat Struct Mol Biol.* 2010;17(7):793–800.
30. Cornish PV, Ermolenko DN, Staple DW, et al. Following movement of the L1 stalk between three functional states in single ribosomes. *Proc Natl Acad Sci USA V106.* 2009; (8):2571–2576.
31. Sternberg SH, Fei J, Prywes N, McGrath KA, Gonzalez Jr RL. Translation factors direct intrinsic ribosome dynamics during translation termination and ribosome recycling. *Nat Struct Mol Biol.* 2009;16(8):861–868.
32. Fei J, Richard AC, Bronson JE, Gonzalez Jr RL. Transfer RNA-mediated regulation of ribosome dynamics during protein synthesis. *Nat Struct Mol Biol.* 2011;18(9):1043–1051.
33. Whitford PC, Geggier P, Altman RB, Blanchard SC, Onuchic JN, Sanbonmatsu KY. Accommodation of aminoacyl-tRNA into the ribosome involves reversible excursions along multiple pathways. *RNA.* 2010;16(6):1196–1204.
34. Blanchard SC, Gonzalez RL, Kim HD, Chu S, Puglisi JD. tRNA selection and kinetic proofreading in translation. *Nat Struct Mol Biol.* 2004;11(10):1008–1014.
35. Ha T, Rasnik I, Cheng W, et al. Initiation and re-initiation of DNA unwinding by the *Escherichia coli* Rep helicase. *Nature.* 2002;419(6907):638–641.
36. Sako Y, Minoghchi S, Yanagida T. Single-molecule imaging of EGFR signalling on the surface of living cells. *Nat Cell Biol.* 2000;2(3):168–172.
37. Moran-Mirabal JM, Craighead HG. Zero-mode waveguides: sub-wavelength nanostructures for single molecule studies at high concentrations. *Methods.* 2008;46(1):11–17.
38. Pedersen S, Bloch PL, Reeh S, Neidhardt FC. Patterns of protein synthesis in *E. coli*: a catalog of the amount of 140 individual proteins at different growth rates. *Cell.* 1978;14 (1):179–190.
39. Howe JG, Hershey JW. A sensitive immunoblotting method for measuring protein synthesis initiation factor levels in lysates of *Escherichia coli*. *J Biol Chem.* 1981;256 (24):12836–12839.
40. Dong H, Nilsson L, Kurland CG. Co-variation of tRNA abundance and codon usage in *Escherichia coli* at different growth rates. *J Mol Biol.* 1996;260(5):649–663.
41. Levene MJ, Korlach J, Turner SW, Foquet M, Craighead HG, Webb WW. Zero-mode waveguides for single-molecule analysis at high concentrations. *Science.* 2003;299 (5607):682–686.
42. Zhu P, Craighead HG. Zero-mode waveguides for single-molecule analysis. *Annu Rev Biophys.* 2012;41:269–293.
43. Samiee KT, Foquet M, Guo L, Cox EC, Craighead HG. lambda-Repressor oligomerization kinetics at high concentrations using fluorescence correlation spectroscopy in zero-mode waveguides. *Biophys J.* 2005;88(3):2145–2153.

44. Foquet M, Samiee KT, Kong X, et al. Improved fabrication of zero-mode waveguides for single-molecule detection. *J Appl Phys.* 2008;103(3):034301.
45. Korlach J, Marks PJ, Cicero RL, et al. Selective aluminum passivation for targeted immobilization of single DNA polymerase molecules in zero-mode waveguide nanostructures. *Proc Natl Acad Sci USA.* 2008;105(4):1176–1181.
46. Lundquist PM, Zhong CF, Zhao P, et al. Parallel confocal detection of single molecules in real time. *Opt Lett.* 2008;33(9):1026–1028.
47. Korlach J, Bibillo A, Wegener J, et al. Long, processive enzymatic DNA synthesis using 100% dye-labeled terminal phosphate-linked nucleotides. *Nucleosides Nucleotides Nucleic Acids.* 2008;27(9):1072–1083.
48. Eid J, Fehr A, Gray J, et al. Real-time DNA sequencing from single polymerase molecules. *Science.* 2009;323(5910):133–138.
49. Moazed D, Noller HF. Intermediate states in the movement of transfer RNA in the ribosome. *Nature.* 1989;342(6246):142–148.
50. Hausner TP, Geigenmuller U, Nierhaus KH. The allosteric three-site model for the ribosomal elongation cycle. New insights into the inhibition mechanisms of aminoglycosides, thiostrepton, and viomycin. *J Biol Chem.* 1988;263(26):13103–13111.
51. Rodnina MV, Wintermeyer W. Fidelity of aminoacyl-tRNA selection on the ribosome: kinetic and structural mechanisms. *Annu Rev Biochem.* 2001;70:415–435.
52. Agirrezabala X, Lei J, Brunelle JL, Ortiz-Meoz RF, Green R, Frank J. Visualization of the hybrid state of tRNA binding promoted by spontaneous ratcheting of the ribosome. *Mol Cell.* 2008;32(2):190–197.
53. Munro JB, Altman RB, O'Connor N, Blanchard SC. Identification of two distinct hybrid state intermediates on the ribosome. *Mol Cell.* 2007;25(4):505–517.
54. Uemura S, Aitken CE, Korlach J, Flusberg BA, Turner SW, Puglisi JD. Real-time tRNA transit on single translating ribosomes at codon resolution. *Nature.* 2010;464(7291):1012–1017.
55. Laursen BS, Sorensen HP, Mortensen KK, Sperling-Petersen HU. Initiation of protein synthesis in bacteria. *Microbiol Mol Biol Rev.* 2005;69(1):101–123.
56. Chen J, Tsai A, O'Leary SE, Petrov A, Puglisi JD. Unraveling the dynamics of ribosome translocation. *Curr Opin Struct Biol.* 2012;22(6):804–814.
57. Tomsic J, Vitali LA, Daviter T, et al. Late events of translation initiation in bacteria: a kinetic analysis. *EMBO J.* 2000;19(9):2127–2136.
58. Milon P, Konevega AL, Gualerzi CO, Rodnina MV. Kinetic checkpoint at a late step in translation initiation. *Mol Cell.* 2008;30(6):712–720.
59. Rodnina MV, Savelsbergh A, Katunin VI, Wintermeyer W. Hydrolysis of GTP by elongation factor G drives tRNA movement on the ribosome. *Nature.* 1997;385(6611):37–41.
60. Fourmy D, Recht MI, Puglisi JD. Binding of neomycin-class aminoglycoside antibiotics to the A-site of 16 S rRNA. *J Mol Biol.* 1998;277(2):347–362.
61. Recht MI, Douthwaite S, Puglisi JD. Basis for prokaryotic specificity of action of aminoglycoside antibiotics. *EMBO J.* 1999;18(11):3133–3138.
62. Shine J, Dalgarno L. The 3′-terminal sequence of *Escherichia coli* 16S ribosomal RNA: complementarity to nonsense triplets and ribosome binding sites. *Proc Natl Acad Sci USA.* 1974;71(4):1342–1346.
63. Calogero RA, Pon CL, Canonaco MA, Gualerzi CO. Selection of the mRNA translation initiation region by *Escherichia coli* ribosomes. *Proc Natl Acad Sci USA.* 1988;85(17):6427–6431.
64. Lockwood AH, Chakraborty PR, Maitra U. A complex between initiation factor IF2, guanosine triphosphate, and fMet-tRNA: an intermediate in initiation complex formation. *Proc Natl Acad Sci USA.* 1971;68(12):3122–3126.
65. Blumberg BM, Nakamoto T, Kezdy FJ. Kinetics of initiation of bacterial protein synthesis. *Proc Natl Acad Sci USA.* 1979;76(1):251–255.

66. Bollen A, Heimark RL, Cozzone A, Traut RR, Hershey JW. Cross-linking of initiation factor IF-2 to *Escherichia coli* 30 S ribosomal proteins with dimethylsuberimidate. *J Biol Chem.* 1975;250(11):4310–4314.
67. Heimark RL, Hershey JW, Traut RR. Cross-linking of initiation factor IF2 to proteins L7/L12 in 70 S ribosomes of *Escherichia coli. J Biol Chem.* 1976;251(24):7779–7784.
68. Caserta E, Tomsic J, Spurio R, La Teana A, Pon CL, Gualerzi CO. Translation initiation factor IF2 interacts with the 30 S ribosomal subunit via two separate binding sites. *J Mol Biol.* 2006;362(4):787–799.
69. Marzi S, Knight W, Brandi L, et al. Ribosomal localization of translation initiation factor IF2. *RNA.* 2003;9(8):958–969.
70. Wintermeyer W, Gualerzi C. Effect of *Escherichia coli* initiation factors on the kinetics of *N*-Acphe-tRNAPhe binding to 30S ribosomal subunits. A fluorescence stopped-flow study. *Biochemistry.* 1983;22(3):690–694.
71. Guenneugues M, Caserta E, Brandi L, et al. Mapping the fMet-tRNA(f)(Met) binding site of initiation factor IF2. *EMBO J.* 2000;19(19):5233–5240.
72. Antoun A, Pavlov MY, Lovmar M, Ehrenberg M. How initiation factors tune the rate of initiation of protein synthesis in bacteria. *EMBO J.* 2006;25(11):2539–2550.
73. Hartz D, McPheeters DS, Gold L. Selection of the initiator tRNA by *Escherichia coli* initiation factors. *Genes Dev.* 1989;3(12A):1899–1912.
74. Grunberg-Manago M, Dessen P, Pantaloni D, Godefroy-Colburn T, Wolfe AD, Dondon J. Light-scattering studies showing the effect of initiation factors on the reversible dissociation of *Escherichia coli* ribosomes. *J Mol Biol.* 1975;94(3):461–478.
75. Dottavio-Martin D, Suttle DP, Ravel JM. The effects of initiation factors IF-1 and IF-3 on the dissociation of *Escherichia coli* 70 S ribosomes. *FEBS Lett.* 1979;97(1):105–110.
76. Dallas A, Noller HF. Interaction of translation initiation factor 3 with the 30S ribosomal subunit. *Mol Cell.* 2001;8(4):855–864.
77. Antoun A, Pavlov MY, Tenson T, Ehrenberg MM. Ribosome formation from subunits studied by stopped-flow and Rayleigh light scattering. *Biol Proced Online.* 2004;6:35–54.
78. Antoun A, Pavlov MY, Lovmar M, Ehrenberg M. How initiation factors maximize the accuracy of tRNA selection in initiation of bacterial protein synthesis. *Mol Cell.* 2006;23 (2):183–193.
79. Milon P, Carotti M, Konevega AL, Wintermeyer W, Rodnina MV, Gualerzi CO. The ribosome-bound initiation factor 2 recruits initiator tRNA to the 30S initiation complex. *EMBO Rep.* 2010;11(4):312–316.
80. Petersen HU, Roll T, Grunberg-Manago M, Clark BF. Specific interaction of initiation factor IF2 of *E. coli* with formylmethionyl-tRNA f Met. *Biochem Biophys Res Commun.* 1979;91(3):1068–1074.
81. Uemura S, Dorywalska M, Lee TH, Kim HD, Puglisi JD, Chu S. Peptide bond formation destabilizes Shine–Dalgarno interaction on the ribosome. *Nature.* 2007;446(7134):454–457.
82. Masuda T, Petrov AN, Iizuka R, Funatsu T, Puglisi JD, Uemura S. Initiation factor 2, tRNA, and 50S subunits cooperatively stabilize mRNAs on the ribosome during initiation. *Proc Natl Acad Sci USA.* 2012;109(13):4881–4885.
83. Tsai A, Petrov A, Marshall RA, Korlach J, Uemura S, Puglisi JD. Heterogeneous pathways and timing of factor departure during translation initiation. *Nature.* 2012;487(7407):390–393.
84. Lockwood AH, Sarkar P, Maitra U. Release of polypeptide chain initiation factor IF-2 during initiation complex formation. *Proc Natl Acad Sci USA.* 1972;69(12):3602–3605.
85. Myasnikov AG, Marzi S, Simonetti A, et al. Conformational transition of initiation factor 2 from the GTP- to GDP-bound state visualized on the ribosome. *Nat Struct Mol Biol.* 2005;12(12):1145–1149.

86. Valle M, Zavialov A, Sengupta J, Rawat U, Ehrenberg M, Frank J. Locking and unlocking of ribosomal motions. *Cell.* 2003;114(1):123–134.
87. Spirin AS. A model of the functioning ribosome: locking and unlocking of the ribosome subparticles. *Cold Spring Harb Symp Quant Biol.* 1969;34:197–207.
88. Frank J, Agrawal RK. A ratchet-like inter-subunit reorganization of the ribosome during translocation. *Nature.* 2000;406(6793):318–322.
89. Zavialov AV, Ehrenberg M. Peptidyl-tRNA regulates the GTPase activity of translation factors. *Cell.* 2003;114(1):113–122.
90. Gavrilova LP, Kostiashkina OE, Koteliansky VE, Rutkevitch NM, Spirin AS. Factor-free ("non-enzymic") and factor-dependent systems of translation of polyuridylic acid by *Escherichia coli* ribosomes. *J Mol Biol.* 1976;101(4):537–552.
91. Jagath JR, Matassova NB, de Leeuw E, et al. Important role of the tetraloop region of 4.5S RNA in SRP binding to its receptor FtsY. *RNA.* 2001;7(2):293–301.
92. Stark H, Rodnina MV, Wieden HJ, van Heel M, Wintermeyer W. Large-scale movement of elongation factor G and extensive conformational change of the ribosome during translocation. *Cell.* 2000;100(3):301–309.
93. Agrawal RK, Heagle AB, Penczek P, Grassucci RA, Frank J. EF-G-dependent GTP hydrolysis induces translocation accompanied by large conformational changes in the 70S ribosome. *Nat Struct Biol.* 1999;6(7):643–647.
94. Wilden B, Savelsbergh A, Rodnina MV, Wintermeyer W. Role and timing of GTP binding and hydrolysis during EF-G-dependent tRNA translocation on the ribosome. *Proc Natl Acad Sci USA.* 2006;103(37):13670–13675.
95. Peske F, Savelsbergh A, Katunin VI, Rodnina MV, Wintermeyer W. Conformational changes of the small ribosomal subunit during elongation factor G-dependent tRNA-mRNA translocation. *J Mol Biol.* 2004;343(5):1183–1194.
96. Hauryliuk V, Mitkevich VA, Eliseeva NA, Petrushanko IY, Ehrenberg M, Makarov AA. The pretranslocation ribosome is targeted by GTP-bound EF-G in partially activated form. *Proc Natl Acad Sci USA.* 2008;105(41):15678–15683.
97. Fei J, Kosuri P, MacDougall DD, Gonzalez Jr RL. Coupling of ribosomal L1 stalk and tRNA dynamics during translation elongation. *Mol Cell.* 2008;30(3):348–359.
98. Fei J, Bronson JE, Hofman JM, Srinivas RL, Wiggins CH, Gonzalez Jr RL. Allosteric collaboration between elongation factor G and the ribosomal L1 stalk directs tRNA movements during translation. *Proc Natl Acad Sci USA.* 2009;106(37):15702–15707.
99. Munro JB, Wasserman MR, Altman RB, Wang L, Blanchard SC. Correlated conformational events in EF-G and the ribosome regulate translocation. *Nat Struct Mol Biol.* 2010;17(12):1470–1477.
100. Chen J, Petrov A, Tsai A, O'Leary SE, Puglisi JD. Coordinated conformational and compositional dynamics drive ribosome translocation. *Nat Struct Mol Biol.* 2013;20(6):718–727.
101. Marshall RA, Dorywalska M, Puglisi JD. Irreversible chemical steps control intersubunit dynamics during translation. *Proc Natl Acad Sci USA.* 2008;105(40):15364–15369.
102. Chen J, Tsai A, Petrov A, Puglisi JD. Nonfluorescent quenchers to correlate single-molecule conformational and compositional dynamics. *J Am Chem Soc.* 2012;134(13):5734–5737.
103. Ermolenko DN, Spiegel PC, Majumdar ZK, Hickerson RP, Clegg RM, Noller HF. The antibiotic viomycin traps the ribosome in an intermediate state of translocation. *Nat Struct Mol Biol.* 2007;14(6):493–497.
104. Stanley RE, Blaha G, Grodzicki RL, Strickler MD, Steitz TA. The structures of the anti-tuberculosis antibiotics viomycin and capreomycin bound to the 70S ribosome. *Nat Struct Mol Biol.* 2010;17(3):289–293.
105. Samiee KT, Moran-Mirabal JM, Cheung YK, Craighead HG. Zero mode waveguides for single-molecule spectroscopy on lipid membranes. *Biophys J.* 2006;90(9):3288–3299.

106. Miyake T, Tanii T, Sonobe H, et al. Real-time imaging of single-molecule fluorescence with a zero-mode waveguide for the analysis of protein–protein interaction. *Anal Chem.* 2008;80(15):6018–6022.
107. Sameshima T, Iizuka R, Ueno T, et al. Single-molecule study on the decay process of the football-shaped GroEL-GroES complex using zero-mode waveguides. *J Biol Chem.* 2010;285(30):23159–23164.
108. Zhao J, Branagan SP, Bohn PW. Single-molecule enzyme dynamics of monomeric sarcosine oxidase in a gold-based zero-mode waveguide. *Appl Spectrosc.* 2012;66(2):163–169.
109. Teng CH, Lionberger TA, Zhang J, Meyhofer E, Ku PC. Fabrication of nanoscale zero-mode waveguides using microlithography for single molecule sensing. *Nanotechnology.* 2012;23(45):455301.
110. Chen J, Dalal RV, Petrov AN, et al. High-throughput platform for real-time monitoring of biological processes by multicolor single-molecule fluorescence. *Proc Natl Acad Sci USA.* 2014;111(2):664–669.
111. Aitken CE, Marshall RA, Puglisi JD. An oxygen scavenging system for improvement of dye stability in single-molecule fluorescence experiments. *Biophys J.* 2008;94(5): 1826–1835.
112. Capece MC, Kornberg GL, Petrov A, Puglisi JD. A simple real-time assay for in vitro translation. *RNA.* 2015;21(2):296–305.
113. Tsai A, Kornberg G, Johansson M, Chen J, Puglisi JD. The dynamics of SecM-induced translational stalling. *Cell Rep.* 2014;7(5):1521–1533.
114. Oliver D, Norman J, Sarker S. Regulation of *Escherichia coli* secA by cellular protein secretion proficiency requires an intact gene X signal sequence and an active translocon. *J Bacteriol.* 1998;180(19):5240–5242.
115. Ito K, Chiba S. Arrest peptides: *cis*-acting modulators of translation. *Annu Rev Biochem.* 2013;82:171–202.
116. Nakatogawa H, Ito K. The ribosomal exit tunnel functions as a discriminating gate. *Cell.* 2002;108(5):629–636.
117. Nakatogawa H, Ito K. Secretion monitor, SecM, undergoes self-translation arrest in the cytosol. *Mol Cell.* 2001;7(1):185–192.
118. Yap MN, Bernstein HD. The plasticity of a translation arrest motif yields insights into nascent polypeptide recognition inside the ribosome tunnel. *Mol Cell.* 2009;34(2): 201–211.
119. Wilson DN, Beckmann R. The ribosomal tunnel as a functional environment for nascent polypeptide folding and translational stalling. *Curr Opin Struct Biol.* 2011;21 (2):274–282.
120. Woolhead CA, Johnson AE, Bernstein HD. Translation arrest requires two-way communication between a nascent polypeptide and the ribosome. *Mol Cell.* 2006;22 (5):587–598.
121. Bhushan S, Hoffmann T, Seidelt B, et al. SecM-stalled ribosomes adopt an altered geometry at the peptidyl transferase center. *PLoS Biol.* 2011;9(1):e1000581.
122. Gumbart J, Schreiner E, Wilson DN, Beckmann R, Schulten K. Mechanisms of Sec M-mediated stalling in the ribosome. *Biophys J.* 2012;103(2):331–341.
123. Pavlov MY, Ehrenberg M. Rate of translation of natural mRNAs in an optimized in vitro system. *Arch Biochem Biophys.* 1996;328(1):9–16.
124. Uemura S, Iizuka R, Ueno T, et al. Single-molecule imaging of full protein synthesis by immobilized ribosomes. *Nucleic Acids Res.* 2008;36(12):e70.
125. Evans MS, Ugrinov KG, Frese MA, Clark PL. Homogeneous stalled ribosome nascent chain complexes produced in vivo or in vitro. *Nat Methods.* 2005;2(10):757–762.
126. Butkus ME, Prundeanu LB, Oliver DB. Translocon "pulling" of nascent SecM controls the duration of its translational pause and secretion-responsive secA regulation. *J Bacteriol.* 2003;185(22):6719–6722.

127. Johansson M, Chen J, Tsai A, Kornberg G, Puglisi JD. Sequence-dependent elongation dynamics on macrolide-bound ribosomes. *Cell Rep.* 2014;7(5):1534–1546.
128. Weisblum B. Insights into erythromycin action from studies of its activity as inducer of resistance. *Antimicrob Agents Chemother.* 1995;39(4):797–805.
129. Menninger JR, Otto DP. Erythromycin, carbomycin, and spiramycin inhibit protein synthesis by stimulating the dissociation of peptidyl-tRNA from ribosomes. *Antimicrob Agents Chemother.* 1982;21(5):811–818.
130. Dunkle JA, Xiong L, Mankin AS, Cate JH. Structures of the *Escherichia coli* ribosome with antibiotics bound near the peptidyl transferase center explain spectra of drug action. *Proc Natl Acad Sci USA.* 2010;107(40):17152–17157.
131. Vazquez-Laslop N, Thum C, Mankin AS. Molecular mechanism of drug-dependent ribosome stalling. *Mol Cell.* 2008;30(2):190–202.
132. Tenson T, Mankin AS. Short peptides conferring resistance to macrolide antibiotics. *Peptides.* 2001;22(10):1661–1668.
133. Lovmar M, Nilsson K, Vimberg V, Tenson T, Nervall M, Ehrenberg M. The molecular mechanism of peptide-mediated erythromycin resistance. *J Biol Chem.* 2006;281(10): 6742–6750.
134. Chen J, Petrov A, Johansson M, Tsai A, O'Leary SE, Puglisi JD. Dynamic pathways of −1 translational frameshifting. *Nature.* 2014;512(7514):328–332.
135. Jenner LB, Demeshkina N, Yusupova G, Yusupov M. Structural aspects of messenger RNA reading frame maintenance by the ribosome. *Nat Struct Mol Biol.* 2010;17 (5):555–560.
136. Tinoco Jr I, Kim HK, Yan S. Frameshifting dynamics. *Biopolymers.* 2013;99(12): 1147–1166.
137. Tsuchihashi Z, Kornberg A. Translational frameshifting generates the gamma subunit of DNA polymerase III holoenzyme. *Proc Natl Acad Sci USA.* 1990;87(7):2516–2520.
138. Larsen B, Gesteland RF, Atkins JF. Structural probing and mutagenic analysis of the stem-loop required for *Escherichia coli* dnaX ribosomal frameshifting: programmed efficiency of 50%. *J Mol Biol.* 1997;271(1):47–60.
139. Larsen B, Wills NM, Gesteland RF, Atkins JF. rRNA–mRNA base pairing stimulates a programmed −1 ribosomal frameshift. *J Bacteriol.* 1994;176(22):6842–6851.
140. Jacks T, Madhani HD, Masiarz FR, Varmus HE. Signals for ribosomal frameshifting in the Rous sarcoma virus gag-pol region. *Cell.* 1988;55(3):447–458.
141. Leger M, Dulude D, Steinberg SV, Brakier-Gingras L. The three transfer RNAs occupying the A, P and E sites on the ribosome are involved in viral programmed −1 ribosomal frameshift. *Nucleic Acids Res.* 2007;35(16):5581–5592.
142. Weiss RB, Dunn DM, Shuh M, Atkins JF, Gesteland RF. *E. coli* ribosomes re-phase on retroviral frameshift signals at rates ranging from 2 to 50 percent. *New Biol.* 1989;1 (2):159–169.
143. Plant EP, Jacobs KL, Harger JW, et al. The 9-A solution: how mRNA pseudoknots promote efficient programmed −1 ribosomal frameshifting. *RNA.* 2003;9(2):168–174.
144. Namy O, Moran SJ, Stuart DI, Gilbert RJ, Brierley I. A mechanical explanation of RNA pseudoknot function in programmed ribosomal frameshifting. *Nature.* 2006;441 (7090):244–247.

CHAPTER TWO

−1 Programmed Ribosomal Frameshifting as a Force-Dependent Process

Koen Visscher
Departments of Physics and Molecular & Cellular Biology, College of Optical Sciences, The University of Arizona, Tucson, Arizona, USA
E-mail address: visscher@email.arizona.edu

Contents

Abstract

−1 Programmed ribosomal frameshifting is a translational recoding event in which ribosomes slip backward along messenger RNA presumably due to increased tension disrupting the codon–anticodon interaction at the ribosome's coding site. Single-molecule physical methods and recent experiments characterizing the physical properties of mRNA's slippery sequence as well as the mechanical stability of downstream mRNA structure motifs that give rise to frameshifting are discussed. Progress in technology, experimental assays, and data analysis methods hold promise for accurate physical modeling and quantitative understanding of −1 programmed ribosomal frameshifting.

Progress in Molecular Biology and Translational Science, Volume 139
ISSN 1877-1173
http://dx.doi.org/10.1016/bs.pmbts.2015.11.003

1. −1 PROGRAMMED RIBOSOMAL FRAMESHIFTING

There is an increasing appreciation for the role of mechanical force in biological processes at the cellular and molecular scale both in vitro and in vivo.[1–5] A striking example that makes it easy to appreciate the role of mechanical force at the molecular level is the translation elongational cycle of the ribosome during protein synthesis, one of life's most fundamental processes. With peptidyl transfer RNA (tRNA) holding on to the growing peptide while residing in the ribosome's P site, aminoacyl transfer RNAs carrying amino acids in accordance with the codon sequence of messenger RNA enter and form a codon–anticodon pair in the aminoacyl or A site of the ribosome (accommodation). After a new peptide bond has formed, the P-site and A-site tRNA's are moved to the E (Exit) site and P site (translocation), respectively, threading paired mRNA through the ribosome. Translocation vacates the A site, allowing new tRNA to enter and bind (Fig. 1). The cycle repeats typically until a stop codon on the mRNA is reached. Since ribosomes can translate only single-stranded mRNA, complex mRNA structure motifs are unwound, by a mechanism that still remains obscure. Unlike initiation of translation for which several helicases have been identified[6] (eIF4A, DHX29, Ded1/DDX3, RHA/DHX9 to name a few), or where ribosomal protein S1 may be involved to unwind the mRNA[7] there are no known RNA helicases associated with the translocation of ribosomes along mRNA as part of the elongational cycle. How then do translocating ribosomes unwind the downstream mRNA structure they encounter? Work by Takyar et al. suggests that prokaryotic ribosomes have helicase activity themselves conferred on them by ribosomal proteins S3, S4, and S5, which encircle the mRNA entering the ribosome.[8] None of these

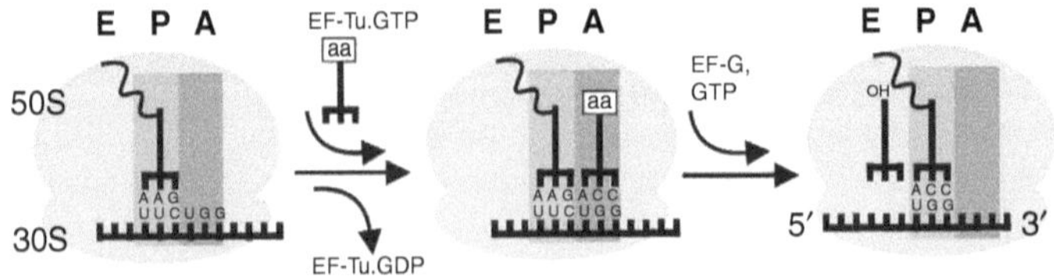

Figure 1 Simplified schematic of the translation elongational cycle. EF-Tu catalyzes the binding of tRNA in the A site, while EF-G catalyzes the translocation step, moving the tRNAs into the P and E sites. *Reprinted with permission from Skinner et al.[15] Copyright 2008 AIP Publishing LLC.*

proteins, however, have any ATPase activity that is usually associated with DNA and/or RNA helicases.[9] However, given the location of the helicase activity inside the mRNA tunnel, rotation of the 30S head relative to the 50S body could serve to shear the mRNA apart.[10,11] It is also in principle conceivable that ribosomes may simply be strong enough, building up sufficient tension when translocating the tRNA and associated mRNA, to unfold any downstream mRNA as it forces it into the narrow mRNA entrance tunnel. However, prokaryotic ribosomes stall out at a relatively low opposing force of approximately 13 pN (for comparison, *Escherichia coli* RNA polymerase can move against forces up to ~24 pN[12]), which seems to rule out any brute-force mechanisms for mRNA unfolding.[13] A recent single-molecule experiment indicates that ribosome helicase action during the translational elongation cycle may be twofold: it destabilizes the helical junction at the mRNA entry site favoring an open conformation, and it appears to pull mRNA strands apart during the translocation step when relatively large structural rearrangements occur on the ribosome.[14] Mechanical force may in principle assist in either activity.

When ribosomes translocate, tRNA and mRNA tension may build up in the single-stranded mRNA connecting the coding site (A and P sites) and any downstream mRNA structure not yet unwound that is being pulled against the mRNA entry tunnel. In response to such increased tension ribosomes may pause or stall; they may unwind the downstream structure; the codon–anticodon interactions at the A and P sites may be disrupted; or a combination of these events may occur. Such a prominent role of mechanical force has appeal for quantitative physical modeling of −1 programmed ribosomal frameshifting (−1 PRF). −1 PRF is a translational recoding event in which ribosomes slip backward, out of the reading frame by a single base in response to two *cis*-acting mRNA elements, a so-called slippery sequence at the ribosome coding site, and a downstream mRNA structure motif (pseudoknot, hairpin) at the entrance to the ribosome (Fig. 2).[16–18] Failure to unwind such a structure increases tension lowering the kinetic barrier for disruption of the codon–anticodon pairs at the slippery site.[19–21] The subsequent mRNA backward slippage, by a single nucleotide, is thought to relieve the tension so that the codon–anticodon interaction is reestablished but now in a −1 shifted reading frame. −1 PRF expands the information capacity of mRNA and is essential for the replication of (retro) viruses with minimal, compact genomes (eg, HIV-1, SARS). In many such systems, −1 PRF induces skipping of the 0-frame stop codon, giving rise to a fusion of structural and enzymatic proteins (to be further edited

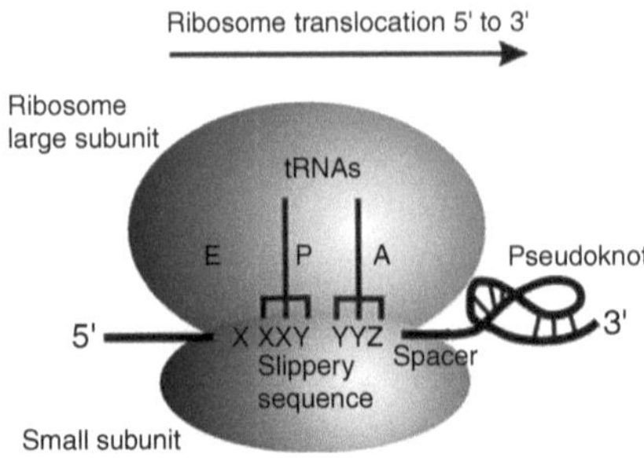

Figure 2 −1 Programmed ribosomal frameshift elements: slippery sequence at the coding site, spacer, and downstream RNA pseudoknot. *Reprinted with permission from Bailey et al.*[24] *Copyright 2014 IOP.*

posttranslationally). In a mechanical picture of −1 PRF, the probability of a frameshift, also known as the frameshift efficiency is determined by the stability of the codon–anticodon interaction at the slippery sequence,[22] the structural integrity of the downstream mRNA structure,[23] and the elastic properties of the spacer bridging both those elements.

When precisely in the translation elongational cycle the −1 frameshift occurs is still up for debate. Plant et al. have hypothesized that even tiny movements as small as 9 Å, thought to occur during the accommodation of tRNA, may increase tension and suffices to disrupt the codon–anticodon pairing when at the slippery sequence.[19] Recent theoretical modeling of −1 PRF shows that the 9 Å hypothesis is physically feasible, showing sharply peaked frameshift efficiencies only when ribosomes are at the slippery site.[24]

While this review addresses how force may be used to characterize elements in translation important for −1 PRF, it is worth pointing out that regulation of translation by force is not exclusively via mRNA, but may also act via other elements, for example, the emerging peptide. Goldman et al., recently demonstrated that mechanical force rescues stalled translation as part of the Sec protein-translocation pathway.[25] This pathway arrests translation as ribosome-nascent peptide chains are targeted to the membrane, that is, to the SecYEG transport channel.[26] Subsequently, the ATPase SecA is thought to physically pull on the nascent peptide restarting translation elongation and driving the peptide though the channel. An elegant single-molecule experiment by Goldman et al. demonstrated that pulling on the peptide with a mechanical force applied by optical tweezers indeed proved sufficient to restart translation elongation in arrested ribosomes.[25]

We will review methods and experiments that apply force to the single molecules to determine the mechanical properties of codon–anticodon interaction at the slippery site, the stability of downstream mRNA structure motifs that give rise to −1 PRF, and elastic properties of the ssRNA elasticity bridging those elements.

2. APPLYING MECHANICAL FORCES TO INDIVIDUAL MOLECULES

While it is next to impossible to apply well-controlled forces to molecules in traditional solution-based biochemical assays,[27] single-molecule techniques such as optical tweezers,[28] magnetic tweezers,[29–31] and atomic force microscopy[32] (AFM)-based force spectroscopy[33] allow forces ranging from sub-pN to the nN range to be applied to individual molecules. Pioneering experiments have included probing the motility of individual motor proteins,[34,35] the stretching of DNA molecules,[36] and the unfolding of giant muscle protein titin,[33] to name a few. These early studies have inspired a wide range of ever more advanced experiments targeting questions about the mechanochemistry of motor proteins[37]; the elastic and structural properties of DNA,[38] RNA,[39] and proteins,[40] as well as the interaction of these molecules with each other or with other ligands (protein-chaperone,[41,42] RNA kissing loops[43,44]).

2.1 Optical Tweezers

In optical tweezers a sharply focused laser beam traps microscopic particles near its focus,[28,45] exerting an approximately linearly increasing force as particles are moved away from the equilibrium position.[46–48] In other words, the optical tweezers act as a microscopic 3-dimensional (3D) spring that may be characterized by 3 spring constant k_x, k_y, and k_z, which are proportional to the laser light intensity used to generate the optical tweezers. In most implementations of optical tweezers, k_x and k_y are similar if not identical, whereas k_z is approximately 3–5 times smaller.[47] A variety of methods, generally based upon the Brownian motion of the particle in a harmonic potential and/or upon the response of the particle due to a known (fluid dynamic) disturbance, have been developed to calibrate these spring constants with ever increasing precision and accuracy.[47,49] Once the spring constants have been determined, forces are computed from the measured displacement of the particle from the center of the tweezers. Optical tweezers have been shown to be ideally suited for exerting and measuring forces in the pN (10^{-12} N) regime. For large displacements the linear force–distance relation is lost and the tweezers is no longer spring-like. In fact a small region of constant force is encountered, which has been taken advantage of for creating a so-called passive constant-force trap.[50] Optical tweezers are not capable of trapping the molecules directly, but rather trap silica or polystyrene microspheres ranging in size from a few hundred nanometers to several micrometers that have been attached to the molecule of interest. To stretch and unfold individual molecules, the structural element of interest (RNA, DNA, protein) usually is sandwiched between two DNA handles that are then attached to the microsphere(s) held with the optical tweezers (Fig. 3).[51,52] Commonly used geometries include a single-beam setup with one handle attached to a solid surface, such as a microscope cover glass, while the other handle is held with the optical tweezers. The molecule may subsequently be stretched by laterally or vertically moving the microscope cover glass using a piezo-driven microscope stage. As indicated in Fig. 3A, due to the specific geometry of this assay, upon lateral movement of the microscope stage the microsphere is pulled and displaced not only laterally but also axially, demanding knowledge of k_x, and k_z as well as measurement of the displacements Δx and Δz to compute the direction and magnitude of

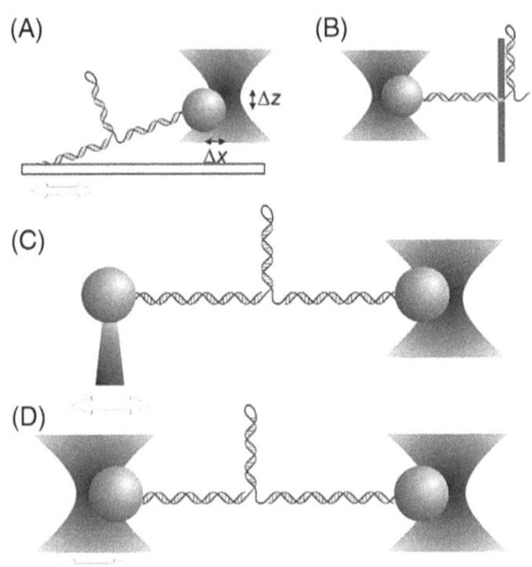

Figure 3 Optical tweezers geometries. (A) Surface tethering. Stage movement induces lateral as well as axial microsphere displacement in the optical tweezers. (B) Optical tweezers-nanopore arrangement. (C) Micropipette-optical tweezers arrangement. (D) Dumbbell arrangement (not to scale).

the force exerted in the direction along the molecule.[53] While initially in these assays the measurement of Δz proved problematic and k_z and Δz were estimated or computed indirectly,[53] it was later realized that the so-called back-focal plane position detection[54,55] provides a direct measurement of Δx, Δy, as well as of Δz with the latter being proportional to the total light intensity impinging on the photosensitive sensor.[56] Recent technological advances have taken advantage of this development and produced ultrastable (low drift) and extremely high precision and resolution optical tweezers specifically for use with the surface-attachment geometry.[57,58]

However, axial displacement in the optical tweezers may be eliminated altogether by avoiding any surface-coupling and alternatively attaching this handle to a microsphere held on a micropipette[51] as indicated in Fig. 3C. Upon proper alignment lateral displacement of the micropipette force measurement then only requires measurement of Δx for accurate computation of force. It is noted that in early implementation of this assay two counter-propagating, moderately focused laser beams were used to form a 3D trap; an arrangement that allows for direct measurement of force (by conservation of momentum).[59]

A preferred assay involves a dumbbell geometry in which the molecule is held in between two optical tweezers (Fig. 3D). This geometry in principle holds significant practical advantages over the other geometries. Instrumental drift is much reduced as the molecule, handles, and beads are suspended in a fluid and do not have any direct mechanical interaction, that is, contact with the instrument. Remaining drift due to laser beam pointing instabilities may be further reduced by traditional-type laser beam stabilization techniques utilizing feedback schemes and/or immersing essential optical paths in helium.[60] Such measures allow for subnanometer, Angstrom level resolution position measurements of trapped particles. In addition, when the positions of the microspheres in both traps are measured simultaneously, correlation analysis may provide further enhancement of the position signal to noise ratio.[61] While in principle not limited to this geometry, the dumbbell assay saw the first implementation of the so-called passive constant-force tweezers or force clamp.[50] Constant-force optical tweezers were pioneered to study the force dependence of the mechanochemistry of the kinesin motor protein.[37,48] This force clamp applied a (digital) feedback to assure a constant distance between the center of the tweezers and the microsphere, thus maintaining a constant force. It has been shown however that the limited time-response of the feedback loop may directly affect the kinetics of unfolding and refolding of molecules held near the transition force.[62] A passive force clamp eliminates the need for any active feedback by pulling one of the microspheres in the dumbbell assay to a region of the optical tweezers where the potential is linear and thus the force is constant.[50] This region typically is small, ~50 nm, limiting its use to the unfolding of small molecules only. Recently by shaping the trapped particles Phillips et al. were able to increase the constant-force region to several microns.[63] It remains to be seen if the relatively large particles for which this was demonstrated prove practical and advantageous for single-molecule unfolding studies, but the idea is interesting as it could allow access to the unfolding of much larger molecules.

With the bandwidth and dead-time limitation of active feedback resolved, folding kinetics may subsequently be derived after correcting the remaining instrument response due to the tweezers' spring constant and compliance of the DNA handles.[64] Constant-force dumbbell assays may also be implemented using counter-propagating laser beams, which allow for shorter than usual DNA handles if so desired.[65,66]

While the emphasis of this review is on the use of optical tweezers for unfolding individual molecules alternative techniques deserve mentioning and in some cases be more beneficial for particular applications.

2.2 Magnetic Tweezers

Magnetic tweezers based assays appear similar to the surface-coupled geometry used with optical tweezers but now the molecule is stretched vertically between the surface and a magnetic bead (Fig. 4A).[30] Two magnets suspended above the bead provide a magnetic field gradient resulting in typical forces in the pN range, its magnitude set by the height of the magnets (or electrical current if electromagnets are used). The typical magnetic field gradient however is sufficiently shallow so that magnetic tweezers are by their nature a constant-force single-molecule technique. The force applied

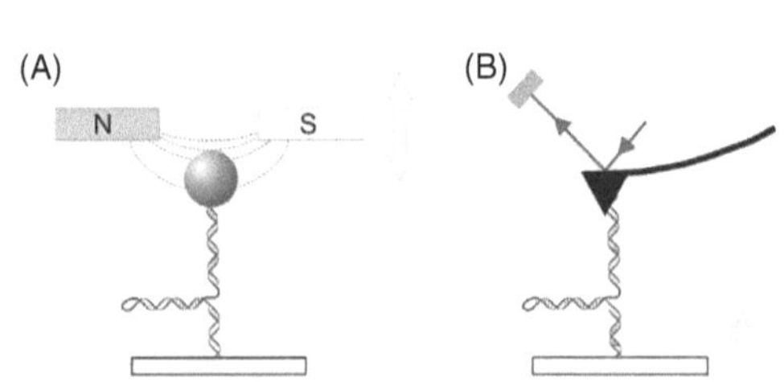

Figure 4 (A) Magnetic tweezers. (B) AFM-based force spectroscopy (not to scale).

remains constant over the dimensions of the molecule, that is, independent of its state, folded or unfolded. As the force is applied vertically, the particle's displacement upon unfolding is usually determined with nanometer resolution from the degree of defocus observed in the microscope using image processing. The axial spring constant, k_z is readily determined from lateral particle displacement fluctuations (Brownian motion) once realizing the system entails an inverted pendulum.[30] It is of interest that axial, vertical stretching is now also used on occasion with optical tweezers enabling the use of short tethers.[67]

While many single-molecule methods readily allow the observation of repeated unfolding and refolding of one particular molecule, sampling a large number of molecules remains time-consuming. However, magnetic tweezers are of particular interest because they hold promise for (massively) parallelization of single-molecule experiments: many surface-coupled beads may readily be observed within one field-of-view of the microscope[68] with all molecules potentially held at the same force. Such parallelization, although in principle not impossible with optical tweezers is more readily achieved using magnetic tweezers in which the magnetic field is applied over a large area that in principle could contain tens to hundreds of surface-tethered particles. It is also noted that recently centrifugation has been reintroduced to exert forces on individual molecules in a massively parallel fashion.[69] However, to date this method does not appear to have been widely adopted.

2.3 AFM

AFM has traditionally been applied for force spectroscopy of individual proteins or protein repeats, in particular the giant protein titin (Fig. 4B).[33] In AFM a cantilevered sharp tip probes the surface which the molecule of interest has been deposited on and/or attached to. As it moves along the surface, the tip approaches and retracts from the surface, detecting a tip-molecule attachment as the cantilever deflects upon retraction. This bending of the cantilever is determined by measuring the deflection of a laser beam reflected off the end of the cantilever using a position-sensitive photo detector. The spring constant of the cantilever is generally determined by analysis of its thermal fluctuations, in much the same way as the optical tweezers use Brownian motion. Other methods exist, for a review see Cumpson et al.[70] While typical cantilevers have spring constants far exceeding those of optical tweezers resulting in typical forces of order nN, softer cantilevers have become available allowing forces in the pN regime to be measured or applied. Furthermore, recent technical improvements reducing drift and increasing tip-sample stability to as small as 100 pm have the potential of bringing AFM to single-molecule unfolding experiments that would otherwise employ optical tweezers.[71] Also note that such stability in principle makes previously impossible constant-force experiments feasible. AFM would then hold the advantage that it is more readily automated to scan the surface for viable candidates for stretching; unfold and refold the molecule a predetermined number of times; then break the tether and move on to the next. Since typically the tip-molecule attachment is nonspecific it is not always clear which part of the molecule is unfolded, or worse if the molecule or something else altogether has been stretched. One therefore tends to construct repeat sequences of the structures of interest which serve as a fingerprint of the molecule.[72] For example, in a five-repeat structure one may decide to only analyze data displaying three or more repeats, thus maximizing the probability that the desired molecule is being unfolded. However, since it was shown that unfolding forces depend upon the number of unfolded domains in a chain, so-called order statistics need to be applied for generating proper parent distribution functions[73] of unfolding forces before such data can further be utilized for analysis and/or predictions of kinetic behavior. While this analysis assumes repeats of identical noninteracting domains governed by a single underlying parent force distribution function, in principle the same applies when unfolding sequences of nonidentical structural domains, such as complex molecules[74] in which case the unfolding forces are pulled from multiple differing parent distributions. Successful statistical order correction may then depend on the ability to recognize which type of domain is unfolded (eg, by difference in change of extension) so as to assign the observed unfolding event to the proper associated parent distribution.

2.4 Nanopores

Biological (eg, α-hemolysin[75,76]) and man-made solid-state[77–79] nanopores enable mechanical unfolding of bare proteins and nucleic acids without the need for any handles or microspheres, thus potentially easing the interpretation of data and derived kinetics.[80] The molecules are forced through a nanometer scale hole due to an electrical potential applied across the pore interrupting the baseline electrical current as the molecule resides in the pore. Unfolding of structures initially too big to pass through and blocking the current may be detected as a sudden increase in the current after they unfold and then zip through the pore. By ramping up the potential difference across the pore, a distribution of unfolding voltages can thus be acquired. Alternatively, one may choose to operate at constant voltage and record the time required for unfolding to occur. In many respects this is analogous to the use of optical tweezers with voltage taking the role of force. However, while nanopore-based experiments sample only a single unfolding event for each of the many molecules passing through sequentially, optical tweezers experiments do the opposite and are better suited to sample many repeated unfolding and refolding events for only one single molecule. Also, unlike optical tweezers the magnitude of the actual force applied remains largely unknown. Efforts have been made to measure this force directly using optical tweezers and nanopores in a combined setup (Fig. 3B).[81] Interestingly, in such an arrangement one could imagine using the optical tweezers to pull a molecule through the nanopore (in the absence of any electrical potential) potentially mimicking how ribosomes pull through and unravel structured mRNA. Although such type of experiments have been proposed[81,82] actual data remains scarce, if nonexistent. For a more extensive discussion see chapter: Studies of RNA Sequence and Structure Using Nanopores.

3. VARYING OR CONSTANT FORCE PROTOCOLS

Molecules may be unfolded following a variety of stretching protocols: one may, for example, slowly or quickly ramp up the force in a linear fashion (force ramp); apply a sudden change in force (force jump); or hold the system at a constant force at all times.[83]

3.1 Force-Ramp Mode

A force ramp is generated simply by moving one of the ends of molecular construct away from the opposing end, usually at a constant speed. As the construct is stretched the force increases accordingly and is determined by measuring the position of the microsphere in the optical trap or magnetic tweezers, or the deflection of the AFM tip. When the force ramp is shallow and the molecule is pulled slowly, that is, slower than the intrinsic kinetic (un)folding rates of the molecule, the (un)folding event occurs at or very near to thermodynamic equilibrium. Equilibrium behavior is readily recognized in recorded force-extension data: the sudden increase in extension upon unfolding during stretching occurs at the same force as the drop in extension upon relaxation of the molecule.[51] In other words, the extension and relaxation curves overlap. At equilibrium conditions, equilibrium thermodynamic quantities such as the free-energy change of the reaction (ΔG) are readily derived once the elastic response of the DNA handles has been accounted for.[51,84,85] However, equilibrium conditions cannot always be met and are often limited to a relatively small group of molecules, such as short DNA and RNA stem loops that lack tertiary interaction and when structural properties are predominantly governed by Watson–Crick base pairing. Nevertheless, even with these molecules, experiments can quickly be driven away from equilibrium by stretching (and relaxing) the molecule more rapidly. In particular, tertiary interactions quickly drive experiments away from equilibrium so that unfolding and refolding forces differ significantly, even at moderate to low ramp rates (an example is shown in Fig. 6). In many cases equilibrium conditions simply prove impractical experimentally as low ramp rates put ever-increasing demands upon instrument stability and drift.

Departure from equilibrium is not as problematic as it may appear. So-called fluctuation relations, such as the Jarzynski's equality and Crooks' fluctuation theorem provide a theoretical framework for deriving equilibrium free energy differences from nonequilibrium work measurements.[86,87] Small RNA hairpins, whose (un)folding using optical tweezers could readily be tuned from near- to off equilibrium by adjusting the force-ramp rate, have successfully served to verify these methods.[88,89] However, applying fluctuation relation may put unexpected demands upon experimental procedures as demonstrated by Ribezzi-Crivellari et al. who point out that the work measurement then should ideally be based upon a force measurement in the moving, not in the static optical tweezers as is customary in dumbbell geometries.[66]

Nonequilibrium measurements also do not preclude the determination of the energy landscapes underlying the physics of molecular folding.[90] Therefore, in some cases it may in fact be advantageous to create off-equilibrium conditions, and to increase ramp rates to ensure low-drift, high-stability experimental conditions.

3.2 Constant-Force Mode

Repeated unfolding and (re)folding events can be recorded for systems at equilibrium by keeping the applied force constant at the characteristic (un) folding force. In the data one would observe the extension to "hop" between two or more discrete mean values representing the folded, unfolded, and possibly intermediate states. For most such experiments, a dumbbell-type passive constant-force optical tweezers is preferred as it circumvents bandwidth issues apparent in active feedback system, and provides a stable, low-drift solution for maintaining a constant force.[50,62] Equilibrium experiments readily allow for the determination of folding energy landscapes. From such experiments, the folding energy landscape $U(x)$ in principle is readily obtained from the histogram or probability distribution of the molecule's extension, $P_{mol}(x)$ as these quantities are related by the Boltzmann distribution at equilibrium[64,90]:

$$P_{\mathrm{mol}}(x) \propto e^{-U(x)/k_{\mathrm{B}}T}$$

However, the real experiment measures $P(x)$, a histogram of microsphere position which also includes contributions due to compliance of the trapped microspheres (the force probe) and the DNA handles, which can be written as a convolution of $P_{mol}(x)$ and a response or point spread function PSF(x) characteristic of the assay used.

$$P(x) = \mathrm{PSF}(x) \otimes P_{\mathrm{mol}}(x),$$

The assay's PSF(x) can be determined by recording microsphere position histograms at identical conditions, but by using the DNA handles microsphere(s) construct without the folded structure motif present. $P_{mol}(x)$ is subsequently computed by applying a (nonlinear) deconvolution algorithm, after which the 1D energy profile readily follows[90]:

$$U(x) \propto -k_{\mathrm{B}}T \ln P_{\mathrm{mol}}(x)$$

The constant-force method relies on the relatively rapid acquisition of a large amount of folding and unfolding events, thus ensuring that the energy barrier separating the folded and unfolded states, where the system spends the least of its time, is also sufficiently sampled. This requirement becomes increasingly hard to meet when the height of the barrier increases and the barrier top becomes rarely sampled by the molecule, leading to impractically long measurement times. To circumvent this problem, de Messieres et al. adopted an umbrella sampling style technique in which a stiff force probes (optical tweezers, AFM) harmonically constrain excursions along the reaction coordinate, but is stepped along the reaction coordinate.[67] As the probe is moved along the reaction coordinate, the otherwise (ie, in constant-force mode) rarely sampled regions are effectively addressed, enabling rapid data acquisition. Landscapes thus reconstructed were in good agreement with those derived from constant-force measurements.

Alternatively one may apply force ramp or jump experiments to sample more closely near (high) energy barriers. However, while the reconstruction of the energy landscape $U(x)$ is rather intuitive at equilibrium conditions, more involved methods, either model independent or model dependent, must be applied to reconstruct energy landscape from nonequilibrium data. Applicability, limitations, advantages, and disadvantages of these approaches have been comprehensively addressed in a recent review by Woodside and Block, and will not be repeated.[90]

4. PROBING −1 PRF ELEMENTS WITH FORCE

4.1 Unfolding RNA Hairpins and Pseudoknots

To theoretically appreciate the effect of force upon mRNA unwinding or unfolding, it is insightful to think of folding as a minimum-energy search on the surface of a landscape reflecting the energetics of the molecular conformations.[91] Such analysis may assist in a quantitative understanding of −1 PRF as it would allow for comparing the thermodynamics and kinetics of the downstream mRNA unfolding with the dynamics and energetics of the disassociation and reassociation of the mRNA codon and tRNA anticodons at the associated slippery sequence. However, given the large number of degrees of freedom of even the smallest RNA molecules, these energy landscapes are highly multidimensional allowing, in principle, multiple pathways from an unfolded to the lowest-energy folded conformation. Experiments usually only resolve a 1D projection of the multidimensional landscape onto a single reaction coordinate (Fig. 5). For example, in the case of mechanical unfolding the end-to-end distance or extension of the molecule appears a natural choice for a reaction coordinate. However, can such a condensed 1D projection capture the relevant thermodynamics and kinetics of the systems studied? In other words, what makes a "good" reaction coordinate? This is an important question as a poor choice of reaction coordinate may quickly lead to incorrect interpretation of data and to wrong conclusions.[90,92] Recent work has started to address this question more rigorously. One may demand that a "good" reaction coordinate faithfully captures the folding kinetics observed in data and provides predictions on the

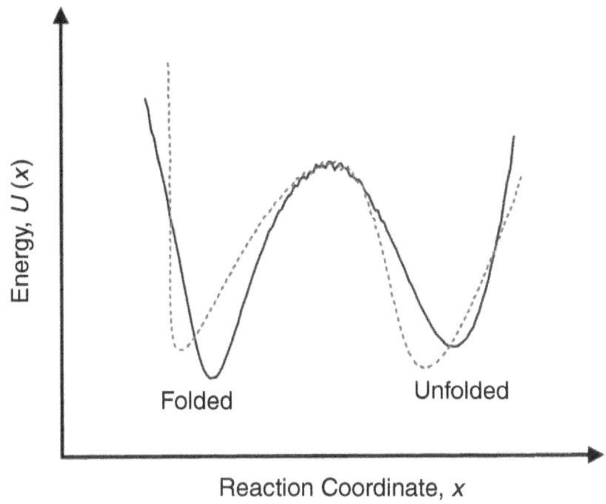

Figure 5 1D energy landscape with a kinetic barrier separating the unfolded and folded states. Application of force tilts the landscape [dashed (red dashed in the web version)] so as to lower the barrier from the folded state promoting escape to the unfolded state.

outcome of a pathway. But since the choice of reaction coordinate is typically not free but experimentally constrained, how can one test and confirm that the experimentally accessible reaction coordinate is in fact a "good" coordinate. Work by Du et al. introduces the split probability or committer, $p_F(x)$, as a reaction coordinate quality test.[93]

With reference to the energy landscape shown in Fig. 5, $p_F(x)$ is defined as the probability along the reaction coordinate x of reaching the folded state. Ideally, $p_F(x)$ should approach 0 when in the unfolded state, 1 in the folded state, and a ½ when at the top of the barrier separating the two energy minima, the latter condition then a reasonable minimum requirement for a "good" reaction coordinate. Upon such analysis by Chodera et al. the end-to-end distance appeared at first a poor reaction coordinate for unfolding DNA hairpins using optical tweezers.[94] However, subsequent work by Neupane et al., who carefully accounted for any effects of the extra fluctuations due to compliance of DNA handles and microspheres in the optical tweezers, did find the end-to-end distance to be a good reaction coordinate.[95] Rather than $p_F(x)$ however, they evaluated a related, perhaps more robust quantity, the conditional probability $p(x|\mathrm{TP})$ that the molecule is on a transition path when the end-to-end distance equals x, a function that sharply peaks near the top of the barrier where it should approach ½. One concludes that (un)folding of simple DNA hairpins near equilibrium can be considered a 1D diffusion problem.[95] While this analysis proved successful for relatively simple molecular structures, it remains to be seen if the same conclusion will hold for more complex architectures containing tertiary interactions prevalent in proteins and many RNA molecules.[74,96] In any case, the analysis framework for testing reaction coordinate quality should be directly applicable to such systems. It may be fair to demand that single-molecule unfolding studies include a reaction coordinate quality test if any energy landscape and/or kinetics are to be derived from single-molecule unfolding data.

As previously discussed, it is not a stretch of the imagination to suppose a correlation between mRNA pseudoknot stability, measured as an unfolding force, and −1 PRF efficiency. However thus far, single-molecule studies of unfolding of −1 PRF-stimulating mRNA structures have predominantly focused on determining unfolding forces with little analyses in terms of energy landscapes and/or kinetics and with little to no attention to the "goodness" of the reaction coordinate. There currently appears somewhat of a disconnect between the advanced analysis methods that have proven successful for small, simple hairpin structures and the unfolding of more complex perhaps more biologically relevant structures.

Stepwise unfolding of large complex RNA structures may provide a fingerprint of its structure when it is dominated by secondary interactions, Watson–Crick base pairing and unfolding proves hierarchical.[74] However, for very short mRNA structures that are dominated by tertiary interactions such fingerprinting is likely to fail. For example, in the case of the Beet Western Yellow Virus RNA pseudoknot (Fig. 6) in which a quadruple base interaction of residue C8 is thought to sustain its compact structure, only the

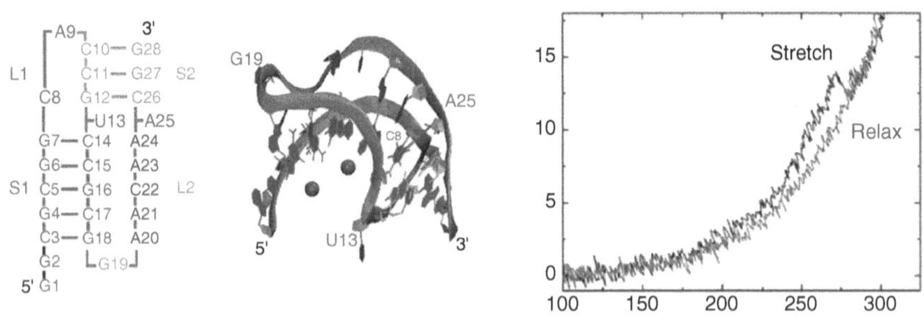

Figure 6 BWYV pseudoknot structures with the C8U substitution mutant's force-extension curves as obtained with surface-mounted optical tweezers in force-ramp mode. Unfolding occurs at a force of ~13 pN, whereas refolding is not observed upon relaxation. *Reprinted with permission from White et al.*[97] *Copyright 2011 American Chemical Society.*

unfolding of stem 1 could explicitly be identified by the increase in extension.[97] Moreover, upon changing pH and/or mutating residue C8 such correlations could no longer be established.

Thus it is not surprising that stretching studies of −1 PRF RNA pseudoknots have thus far focused on determining the forces required for unfolding wild-type and mutant structures that have increased or reduced frameshift efficiency. While early stretching studies indicate that a correlation between frameshift efficiencies and unfolding force exists,[98,99] later studies fail to establish any such correlation.[97,100,101] Interestingly, Ritchie et al. propose a correlation with conformational plasticity of the pseudoknot rather than with force, consistent with White's observation that BWYV mutant pseudoknots abolishing −1 PRF in fact yield slightly higher unfolding forces and may be accessing multiple unfolding pathways much unlike the wild-type BWYV pseudoknot.[97,101] Furthermore, −1 PRF-suppressing ligands also appear to inhibit such required structural plasticity in the SARS coronavirus pseudoknot.[102] A recent single-molecule analysis of the nonviral, human −1 PRF signal in CCR5 mRNA also has yielded multiple conformers and unfolding pathways.[103]

It is hard to judge how appropriate and insightful these unfolding experiments truly are for a quantitative physical understanding of −1 PRF. During the translational elongation cycle, the mRNA, rather than being stretched from both ends, is threaded through the narrow entry tunnel to the ribosome, leaving one end free. Therefore, both scenarios may access significantly different unfolding pathways on a multidimensional energy landscape. Such concern may suggest alternative experimental geometries in which individual RNA molecules are pulled through nanopores, ideally in a controlled fashion using optical tweezers to apply well-defined forces.[82]

Unfolding by optical tweezers, or as proposed using nanopores are, with "bare" mRNA molecules, that is, in the absence of any upstream ribosomes. It will be of interest to see if ribosome-positioned upstream significantly affect pseudoknot unfolding force, unfolding pathways, and plasticity.[104,105] Single-molecule assays that probe the codon–anticodon dynamics at the slippery sequence and investigate the force dependence of the translation elongational cycle (discussed later in the chapter; see Fig. 8) should be fit to answer these questions and allow unfolding of −1 PRF mRNA structure motifs held at the ribosome's entry site.[13,106] Irrespective of such concerns, unfolding of isolated molecules may still provide clues toward a −1 PRF mechanism in specific cases. For example, in the case of the BWYV pseudoknot, combining unfolding with optical tweezers and computationally by steered molecular dynamics simulations, identified detailed molecular rearrangements that may explain how the C8U and C8A mutations abolish −1 PRF.[97]

4.2 Elastic Properties of the Single-Stranded RNA Spacer Sequence

Connecting the slippery sequence at the coding site and the downstream mRNA structure motif is a short (~5–6 nucleotides) single-stranded spacer sequence. How rapidly tension builds up upon ribosome motion or is relieved upon mRNA unfolding, depends upon the spacer's elastic properties. Due to lack of any other data, quantitative models usually apply elastic properties as derived from single-molecule stretching experiments using long, spatially unconfined RNA molecules, often homopolymeric to avoid any base pairing.[107–109]

Both AFM and optical tweezers have been used for such single-molecule stretching experiments, which allowed the molecule to be modeled as a so-called worm-like chain model for polymer elasticity, yielding a persistence length, that is, a measure for its stiffness expressed as the typical distance over which a thermal energy unit of $k_B T$ is able to bend the molecule, of approximately 1–1.5 nm, about 30–50 times smaller than for dsDNA. The worm-like chain's force–extension relation then allows computation of the developed tension[53]:

$$F(x) = \frac{k_B T}{L_p}\left[\frac{1}{4(1 - x/L + F/S)^2} - \frac{1}{4} + \frac{x}{L} - \frac{F}{S}\right],$$

with k_B the Boltzmann constant, L_p the persistence length, L the contour length of the molecule, and S the stretching modulus. Examples of force-extension data for homopolymeric single-stranded RNA are shown in Fig. 7.

However, it is not obvious that such data and associated polymer elasticity models are applicable to the short, heteropolymeric and spatially confined spacer sequence. Further complications arise when considering that the spacer is likely to interact with ribosomal RNA and proteins lining the tunnel,[11,110] something not readily captured by any polymer model. In a stochastic model of −1 PRF, Bailey et al. found it necessary to account for the effects of confinement,[24] which proved somewhat problematic as theories either address very strong (channels much narrower than a persistence length) or weak (channels much wider than a persistence length) confinement.[111,112] Recent work such as by Tree et al. explores the intermediate

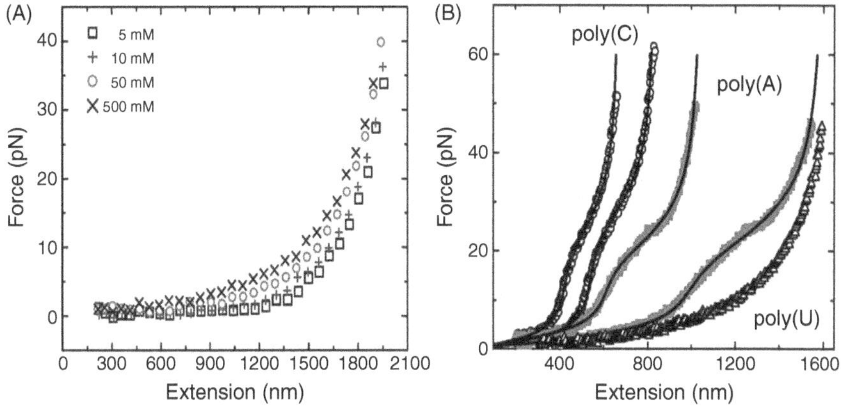

Figure 7 Elasticity of RNA homopolymers. (A) Force-extension of a single poly(U) RNA molecule at varying Na^+ concentrations. A worm-like chain model fits the 500 mM data well but fails to capture the increasingly prevalent electrostatic interactions at low Na^+ concentrations.[107] (B) Unlike poly(U) the force-extension data for poly(A) and poly(C) displays a shoulder indicative of an helix-coil transition. Solid lines fit to a theoretical model.[109] *Reprinted with permission from Seol et al.*[107,109] *Copyright 2004, 2007 American Physical Society.*

regime of confinement of more direct relevance to the spacer sequence.[113] However, such work often does not address the force–extension relation under moderately confined conditions[114] which would be beneficial for modeling and a quantitative understanding of −1 PRF.[24]

4.3 Probing the Coding Site

What remains is a physical characterization of codon–anticodon interactions, ideally with slippery as well as nonslippery mRNA sequences at the ribosome coding site.

Early work by Uemura et al. report rupture forces for mRNA and prokaryotic ribosomes of 12–15 pN in the absence of a flanking Shine–Dalgarno sequence, but with the A and P sites occupied with tRNAs.[115] Addition of an upstream Shine–Dalgarno sequence stabilizes the mRNA–ribosome complex yielding rupture forces as high as ~27 pN. Vanzi et al. reported sudden slippage of ribosomes on a poly(U) template at typical forces of 10–15 pN.[108] However, active translocation was not demonstrated in these studies. At conditions favorable for active translocation, Skinner[15] found gradual slippage on poly(U) at forces as low as 7–9 pN. These measurements may serve to provide typical baseline force values to be kept in mind in the context of frameshifting.

Recent experiments by the Bustamante and Tinoco groups address translation by active prokaryotic ribosomes at the single-molecule level,[13,14,106,116] but do not yet synthesize a complete answer of how force affects the codon–anticodon interaction and its dynamics on the ribosome during frameshifting. But, since these experimental assays together appear to hold all the ingredients to successfully address this question the experimental geometry and some of their results are outlined later in the chapter. An ingenious single-molecule experiment by Yan et al. primarily probes the effect of the slippery sequence and flanking mRNA structure motifs on codon–anticodon interaction and −1 frameshift efficiency[106] (Fig. 8), whereas Liu et al. explicitly explore the role of force upon translation velocities on nonslippery mRNA sequences.[13]

In Yan's setup the mRNA being translated, containing the frameshift signal is designed as a hairpin sandwiched between two DNA handles with their ends held in optical tweezers and on the tip of a micropipette, respectively (Fig. 8). When during translation, ribosomes move along and unwind the double-stranded hairpin, upon each translated codon six nucleotides

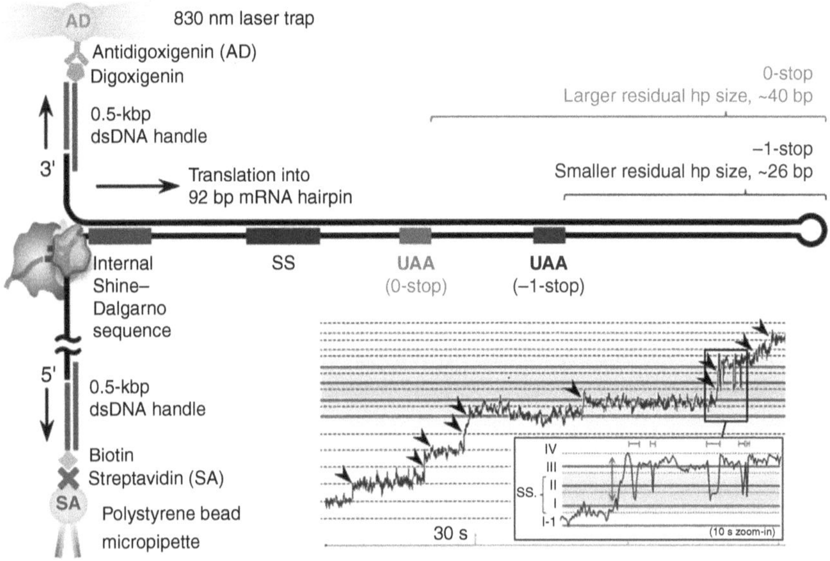

Figure 8 Single-molecule assay for detecting codon-sampling (fluctuations) and −1 frameshifting by individual ribosomes at the slippery sequence (SS). Dotted lines are spaced one codon apart. *Reprinted with permission from Yan et al.[106] Copyright 2015 Elsevier.*

become exposed yielding a readily detectable stepwise increase in extension of the trapped DNA–RNA construct. Downstream of the frameshift signal the mRNA contains two spatially well-separated stop codons positioned in the 0 frame and −1 frame, respectively, which serve as frameshift reporters (Fig. 8). Once a ribosome has stalled at one of the stop codons, a long (∼40 bp) or short (∼26 bp) hairpin remains depending upon the reading frame. The size of the remaining hairpin is readily determined by unfolding using the optical tweezers in force-ramp mode and careful analysis of the increase in extension. The frameshift signal investigated was derived from the *E. coli DnaX* gene and contains three essential RNA elements: a slippery sequence (AAAAAAC) and a downstream RNA structure (a 11 base pair hairpin), two elements also commonly required in viral −1 PRF signals, and in addition a Shine–Dalgarno sequence 10 nucleotides upstream of the slippery sequence. In the assay the *DnaX* hairpin has been extended to contain the frameshift reporters, that is, the two stop codons and hairpin structure.

Using this assay the authors observe stepwise (one codon at a time) unidirectional movement of individual ribosomes along mRNA. However, upon reaching the particular sequence—slippery or not—flanked by the Shine–Dalgarno sequence and the downstream hairpin junction, dynamic (∼2 Hz), large scale (∼1 codon) back-and-forth fluctuations were observed (Fig. 8). These fluctuations are thought to allow the ribosome to sample the different reading frames (codon sampling). Interestingly, smaller but higher-frequency fluctuations at, for example, 85 Hz as identified by power spectral analysis occur exclusively on the flanked slippery sequence (wild-type and A5G mutant) possibly reflecting ribosome subunit head rotation enhanced by the flanking structural elements. By combining mass spectrometry analysis of translated product and single-molecule translation trajectories the authors identify a wide range of frameshifting pathways, and also find evidence for fidelity checking of codon–anticodon that gives rise to early termination after frameshifting.

In Yan's single-molecule frameshift assay the ribosome has not been conjugated to any surface or microsphere and is free to move along the mRNA threaded through it. This renders it less well suited for probing the force dependence of the observed phenomena, such as the codon sampling, the low- and high-frequency fluctuations, and frameshift efficiency. In the current configuration the entire mRNA–DNA handle complex was held at a constant force of ∼18 pN. As the mRNA path through the ribosome is not straight, any tension trying to straighten the mRNA may in principle reduce accessible mRNA conformations required at the coding site and indirectly destabilize any tRNA–mRNA interactions. However, such concerns appear unwarranted as previous work did not find any tension dependence of translocation step times.[116] In fact, high tension proved beneficial as it rescued long ribosomal stalls, presumably by assisting in unfolding any downstream mRNA roadblocks.

To more directly probe the effect of force at the coding site, the ribosome itself needs to be immobilized, while force pulls on the mRNA.[13,15,108,115] Liu et al. have used such an experimental geometry and measured a single-exponential force dependence of the translational velocity. Surface attachment was via ribosomal protein S16, at point thought to minimally affect ribosome function so that force exerted by the mRNA would predominantly affect the mechanical translocation of the tRNAs from the A and P sites to the P and E site, respectively. A stall force of 13 ± 2 pN was found (by extrapolation), lower than, for example, the unfolding force of stemloop 1 of the BWYV pseudoknot which promotes frameshifting efficiently in prokaryotic translation system.[97,105] This goes to show that downstream mRNA structure motifs can readily regulate translation velocities and therefore may affect associated phenomena such as frameshifting. Also, since the actual applied force did not appear to exceed 10 pN, no mRNA–ribosome ruptures were reported.[108,115]

It should be clear that these assays are uniquely qualified to combine the direct force measurements à la Liu[13] with the reading frameshift reporting mRNA template developed by Yan et al.[106] If then also the mRNA template contains a slippery sequence associated with a mechanically (ie, by force) well-characterized −1 PRF-promoting hairpin or pseudoknot, a complete physical picture of −1 PRF may successfully emerge.

It is noted that while −1 PRF is mainly a eukaryotic phenomenon, all single-molecule translation assays employ prokaryotic ribosomes. However, it has also been shown that −1 PRF systems can frameshift efficiently in prokaryotic translation assays. Furthermore while the biological details may differ, it is not unlikely that the underlying physical principles remain the same.[105,110]

5. CONCLUSIONS

Ever-advancing physical methods and single-molecule assay designs combined with sophisticated data-analysis methods bring a precise physical mechanism and model of −1 PRF within reach. Although mechanical force has a dominant role to play, the merging picture will likely be rather more complex than any assumed simple correlation of −1 PRF efficiency and RNA pseudoknot unfolding force. How force controls the kinetics of codon sampling at the slippery sequence and of the melting of the downstream mRNA structure is the key to a quantitative model of −1 PRF.

REFERENCES

1. Yu H, Mouw JK, Weaver VM. Forcing form and function: biomechanical regulation of tumor evolution. *Trends Cell Biol.* 2011;21(1):47–56.
2. Diz-Muñoz A, Fletcher DA, Weiner OD. Use the force: membrane tension as an organizer of cell shape and motility. *Trends Cell Biol.* 2013;23(2):47–53.
3. Iskratsch T, Wolfenson H, Sheetz MP. Appreciating force and shape—the rise of mechanotransduction in cell biology. *Nat Rev Mol Cell Biol.* 2014;15(12):825–833.
4. Tinoco I. Force as a useful variable in reactions: unfolding RNA. *Annu Rev Biophys Biomol Struct.* 2004;33:363–385.
5. Roberts AJ, Kon T, Knight PJ, Sutoh K, Burgess SA. Functions and mechanics of dynein motor proteins. *Nat Rev Mol Cell Biol.* 2013;14(11):713–726.
6. Marintchev A. Roles of helicases in translation initiation: a mechanistic view. *Biochim Biophys Acta.* 2013;1829(8):799–809.
7. Qu X, Lancaster L, Noller HF, Bustamante C, Tinoco I. Ribosomal protein S1 unwinds double-stranded RNA in multiple steps. *Proc Natl Acad Sci.* 2012;109(36):14458–14463.
8. Takyar S, Hickerson RP, Noller HF. mRNA helicase activity of the ribosome. *Cell.* 2005;120(1):49–58.
9. von Hippel PH, Delagoutte E. A general model for nucleic acid helicases and their "coupling" within macromolecular machines. *Cell.* 2001;104(2):177–190.
10. Agrawal RK, Heagle AB, Penczek P, Grassucci RA, Frank J. EF-G-dependent GTP hydrolysis induces translocation accompanied by large conformational changes in the 70S ribosome. *Nat Struct Biol.* 1999;6(7):643–647.
11. Yusupova GZ, Yusupov MM, Cate JH, Noller HF. The path of messenger RNA through the ribosome. *Cell.* 2001;106(2):233–241.
12. Wang MD, Schnitzer MJ, Yin H, Landick R, Gelles J, Block SM. Force and velocity measured for single molecules of RNA polymerase. *Science.* 1998;282(5390):902.
13. Liu T, Kaplan A, Alexander L, et al. Direct measurement of the mechanical work during translocation by the ribosome. *eLife.* 2014;3.
14. Qu X, Wen J-D, Lancaster L, Noller HF, Bustamante C, Tinoco I. The ribosome uses two active mechanisms to unwind messenger RNA during translation. *Nature.* 2011;475(7354):118–121.
15. Skinner GM, Seol Y, Visscher K, Dagdug L, Gracía-Colin SL. Translation against an applied force. In: *AIP Conference Proceedings*; 2008:1–10
16. Jacks T, Power MD, Masiarz FR, Luciw PA, Barr PJ, Varmus HE. Characterization of ribosomal frameshifting in HIV-1 gag-pol expression. *Nature.* 1988;331(6153):280–283.
17. Parkin NT, Chamorro M, Varmus HE. Human immunodeficiency virus type 1 gag-pol frameshifting is dependent on downstream mRNA secondary structure: demonstration by expression in vivo. *J Virol.* 1992;66(8):5147–5151.
18. Giedroc DP, Cornish PV. Frameshifting RNA pseudoknots: structure and mechanism. *Virus Res.* 2009;139(2):193–208.
19. Plant EP, Jacobs KLM, Harger JW, et al. The 9-A solution: how mRNA pseudoknots promote efficient programmed −1 ribosomal frameshifting. *RNA.* 2003;9(2):168–174.

20. Plant EP, Dinman JD. Torsional restraint: a new twist on frameshifting pseudoknots. *Nucleic Acids Res.* 2005;33(6):1825–1833.
21. Namy O, Moran SJ, Stuart DI, Gilbert RJC, Brierley I. A mechanical explanation of RNA pseudoknot function in programmed ribosomal frameshifting. *Nature.* 2006;441 (7090):244–247.
22. Brierley I, Jenner AJ, Inglis SC. Mutational analysis of the "slippery-sequence" component of a coronavirus ribosomal frameshifting signal. *J Mol Biol.* 1992;227(2): 463–479.
23. Kim YG, Su L, Maas S, O'Neill A, Rich A. Specific mutations in a viral RNA pseudoknot drastically change ribosomal frameshifting efficiency. *Proc Natl Acad Sci USA.* 1999;96(25):14234–14239.
24. Bailey BL, Visscher K, Watkins J. A stochastic model of translation with −1 programmed ribosomal frameshifting. *Phys Biol.* 2014;11(1):016009.
25. Goldman DH, Kaiser CM, Milin A, Righini M, Tinoco I, Bustamante C. Mechanical force releases nascent chain-mediated ribosome arrest in vitro and in vivo. *Science.* 2015;348(6233):457–460.
26. Mori H, Ito K. The Sec protein-translocation pathway. *Trends Microbiol.* 2001;9 (10):494–500.
27. Hall K, Cole D, Yeh Y, Baskin RJ. Kinesin force generation measured using a centrifuge microscope sperm-gliding motility assay. *Biophys J.* 1996;71(6):3467–3476.
28. Ashkin A, Dziedzic JM, Bjorkholm JE, Chu S. Observation of a single-beam gradient force optical trap for dielectric particles. *Opt Lett.* 1986;11(5):288.
29. Strick TR, Allemand J-F, Bensimon D, Bensimon A, Croquette V. The elasticity of a single supercoiled DNA molecule. *Science.* 1996;271(5257):1835–1837.
30. Gosse C, Croquette V. Magnetic tweezers: micromanipulation and force measurement at the molecular level. *Biophys J.* 2002;82(6):3314–3329.
31. De Vlaminck I, Dekker C. Recent advances in magnetic tweezers. *Annu Rev Biophys.* 2012;41(1):453–472.
32. Binnig G, Quate CF, Gerber C. Atomic force microscope. *Phys Rev Lett.* 1986;56 (9):930–933.
33. Rief M. Reversible unfolding of individual titin immunoglobulin domains by AFM. *Science.* 1997;276(5315):1109–1112.
34. Block SM, Goldstein LSB, Schnapp BJ. Bead movement by single kinesin molecules studied with optical tweezers. *Nature.* 1990;348(6299):348–352.
35. Svoboda K, Schmidt CF, Schnapp BJ, Block SM. Direct observation of kinesin stepping by optical trapping interferometry. *Nature.* 1993;365(6448):721–727.
36. Smith S, Finzi L, Bustamante C. Direct mechanical measurements of the elasticity of single DNA molecules by using magnetic beads. *Science.* 1992;258(5085):1122–1126.
37. Visscher K, Schnitzer MJ, Block SM. Single kinesin molecules studied with a molecular force clamp. *Nature.* 1999;400(6740):184–189.
38. Bustamante C, Smith SB, Liphardt J, Smith D. Single-molecule studies of DNA mechanics. *Curr Opin Struct Biol.* 2000;10(3):279–285.
39. Lipfert J, Skinner GM, Keegstra JM, et al. Double-stranded RNA under force and torque: similarities to and striking differences from double-stranded DNA. *Proc Natl Acad Sci USA.* 2014;111(43):15408–15413.
40. Chen H, Yuan G, Winardhi RS, et al. Dynamics of equilibrium folding and unfolding transitions of titin immunoglobulin domain under constant forces. *J Am Chem Soc.* 2015;137(10):3540–3546.
41. Bechtluft P, van Leeuwen RGH, Tyreman M, et al. Direct observation of chaperone-induced changes in a protein folding pathway. *Science.* 2007;318(5855):1458–1461.
42. Mashaghi A, Kramer G, Lamb DC, Mayer MP, Tans SJ. Chaperone action at the single-molecule level. *Chem Rev.* 2014;114(1):660–676.

43. Li PTX, Bustamante C, Tinoco I. Unusual mechanical stability of a minimal RNA kissing complex. *Proc Natl Acad Sci USA.* 2006;103(43):15847–15852.
44. Li PTX, Tinoco I. Mechanical unfolding of two DIS RNA kissing complexes from HIV-1. *J Mol Biol.* 2009;386(5):1343–1356.
45. Ashkin A. Optical trapping and manipulation of neutral particles using lasers. *Proc Natl Acad Sci USA.* 1997;94(10):4853.
46. Block S, Svoboda K. Biological applications of optical forces. *Annu Rev Biophys Biomol Struct.* 1994;23:247–285.
47. Neuman KC, Block SM. Optical trapping. *Rev Sci Instrum.* 2004;75(9):2787–2809.
48. Visscher K, Block SM. Versatile optical traps with feedback control. *Methods Enzymol.* 1998;298:460–489.
49. Neuman KC, Nagy A. Single-molecule force spectroscopy: optical tweezers, magnetic tweezers and atomic force microscopy. *Nat Methods.* 2008;5(6):491–505.
50. Greenleaf W, Woodside M, Abbondanzieri E, Block S. Passive all-optical force clamp for high-resolution laser trapping. *Phys Rev Lett.* 2005;95(20):208102.
51. Liphardt J, Onoa B, Smith SB, Tinoco Jr I, Bustamante C. Reversible unfolding of single RNA molecules by mechanical force. *Science.* 2001;292(5517):733.
52. Cecconi C. Direct observation of the three-state folding of a single protein molecule. *Science.* 2005;309(5743):2057–2060.
53. Wang MD, Yin H, Landick R, Gelles J, Block SM. Stretching DNA with optical tweezers. *Biophys J.* 1997;72(3):1335–1346.
54. Visscher K, Block SM. Versatile optical traps with feedback control. In: *Methods in Enzymology.* Vol 298. Elsevier; 1998:460–489. http://linkinghub.elsevier.com/retrieve/pii/S0076687998980405
55. Gittes F, Schmidt CF. Interference model for back-focal-plane displacement detection in optical tweezers. *Opt Lett.* 1998;23(1):7.
56. Rohrbach A, Stelzer EHK. Three-dimensional position detection of optically trapped dielectric particles. *J Appl Phys.* 2002;91(8):5474.
57. Perkins TT. Ångström-precision optical traps and applications. *Annu Rev Biophys.* 2014;43(1):279–302.
58. Walder R, Paik DH, Bull MS, Sauer C, Perkins TT. Ultrastable measurement platform: sub-nm drift over hours in 3D at room temperature. *Opt Express.* 2015;23(13): 16554–16564.
59. Smith SB, Cui Y, Bustamante C. Overstretching B-DNA: the elastic response of individual double-stranded and single-stranded DNA molecules. *Science.* 1996;271 (5250):795–799.
60. Abbondanzieri EA, Greenleaf WJ, Shaevitz JW, Landick R, Block SM. Direct observation of base-pair stepping by RNA polymerase. *Nature.* 2005;438(7067):460–465.
61. Moffitt JR, Chemla YR, Izhaky D, Bustamante C. Differential detection of dual traps improves the spatial resolution of optical tweezers. *Proc Natl Acad Sci USA.* 2006;103 (24):9006–9011.
62. Elms PJ, Chodera JD, Bustamante CJ, Marqusee S. Limitations of constant-force-feedback experiments. *Biophys J.* 2012;103(7):1490–1499.
63. Phillips DB, Padgett MJ, Hanna S, et al. Shape-induced force fields in optical trapping. *Nat Photon.* 2014;8(5):400–405.
64. Woodside MT, Anthony PC, Behnke-Parks WM, Larizadeh K, Herschlag D, Block SM. Direct measurement of the full, sequence-dependent folding landscape of a nucleic acid. *Science.* 2006;314(5801):1001–1004.
65. Ribezzi-Crivellari M, Huguet JM, Ritort F. Counter-propagating dual-trap optical tweezers based on linear momentum conservation. *Rev Sci Instrum.* 2013;84(4):043104.
66. Ribezzi-Crivellari M, Ritort F. Free-energy inference from partial work measurements in small systems. *Proc Natl Acad Sci USA.* 2014;111(33):E3386–E3394.

67. de Messieres M, Brawn-Cinani B, La Porta A. Measuring the folding landscape of a harmonically constrained biopolymer. *Biophys J.* 2011;100(11):2736–2744.
68. Dulin D, Vilfan ID, Berghuis BA, et al. Elongation-competent pauses govern the fidelity of a viral RNA-dependent RNA polymerase. *Cell Rep.* 2015;10(6):983–992.
69. Halvorsen K, Wong WP. Massively parallel single-molecule manipulation using centrifugal force. *Biophys J.* 2010;98(11):L53–L55.
70. Cumpson PJ, Clifford CA, Portoles JF, Johnstone JE, Munz M. Cantilever spring-constant calibration in atomic force microscopy. In: Bhushan B, Fuchs H, Tomitori M, eds. In: *Applied Scanning Probe Methods VIII. Berlin, Heidelberg: Springer Berlin Heidelberg; 2008:. http://link.springer.com/10.1007/978-3-540-74080-3_8*
71. Churnside AB, Perkins TT. Ultrastable atomic force microscopy: improved force and positional stability. *FEBS Lett.* 2014;588(19):3621–3630.
72. Li L, Wetzel S, Plückthun A, Fernandez JM. Stepwise unfolding of ankyrin repeats in a single protein revealed by atomic force microscopy. *Biophys J.* 2006;90(4):L30–L32.
73. Zhmurov A, Dima RI, Barsegov V. Order statistics theory of unfolding of multimeric proteins. *Biophys J.* 2010;99(6):1959–1968.
74. Harlepp S, Marchal T, Robert J, et al. Probing complex RNA structures by mechanical force. *Eur Phys J E Soft Matter.* 2003;12(4):605–615.
75. Akeson M, Branton D, Kasianowicz JJ, Brandin E, Deamer DW. Microsecond time-scale discrimination among polycytidylic acid, polyadenylic acid, and polyuridylic acid as homopolymers or as segments within single RNA molecules. *Biophys J.* 1999;77 (6):3227–3233.
76. Meller A, Nivon L, Brandin E, Golovchenko J, Branton D. Rapid nanopore discrimination between single polynucleotide molecules. *Proc Natl Acad Sci USA.* 2000;97 (3):1079–1084.
77. Li J, Stein D, McMullan C, Branton D, Aziz MJ, Golovchenko JA. Ion-beam sculpting at nanometre length scales. *Nature.* 2001;412(6843):166–169.
78. Storm AJ, Chen JH, Ling XS, Zandbergen HW, Dekker C. Fabrication of solid-state nanopores with single-nanometre precision. *Nat Mater.* 2003;2(8):537–540.
79. Li J, Gershow M, Stein D, Brandin E, Golovchenko JA. DNA molecules and configurations in a solid-state nanopore microscope. *Nat Mater.* 2003;2(9):611–615.
80. Dudko OK, Mathé J, Szabo A, Meller A, Hummer G. Extracting kinetics from single-molecule force spectroscopy: nanopore unzipping of DNA hairpins. *Biophys J.* 2007;92 (12):4188–4195.
81. van den Hout M, Vilfan ID, Hage S, Dekker NH. Direct force measurements on double-stranded RNA in solid-state nanopores. *Nano Lett.* 2010;10(2):701–707.
82. Dekker C. Solid-state nanopores. *Nat Nanotechnol.* 2007;2(4):209–215.
83. Tinoco I, Li PTX, Bustamante C. Determination of thermodynamics and kinetics of RNA reactions by force. *Q Rev Biophys.* 2006;39(04):325.
84. Wen J-D, Manosas M, Li PTX, et al. Force unfolding kinetics of RNA using optical tweezers. I. Effects of experimental variables on measured results. *Biophys J.* 2007;92 (9):2996–3009.
85. Manosas M, Wen J-D, Li PTX, et al. Force unfolding kinetics of RNA using optical tweezers. II. Modeling experiments. *Biophys J.* 2007;92(9):3010–3021.
86. Jarzynski C. Nonequilibrium equality for free energy differences. *Phys Rev Lett.* 1997;78 (14):2690–2693.
87. Crooks GE. Entropy production fluctuation theorem and the nonequilibrium work relation for free energy differences. *Phys Rev E.* 1999;60(3):2721–2726.
88. Liphardt J, Dumont S, Smith SB, Tinoco I, Bustamante C. Equilibrium information from nonequilibrium measurements in an experimental test of Jarzynski's equality. *Science.* 2002;296(5574):1832–1835.

89. Collin D, Ritort F, Jarzynski C, Smith SB, Tinoco I, Bustamante C. Verification of the Crooks fluctuation theorem and recovery of RNA folding free energies. *Nature.* 2005;437(7056):231–234.
90. Woodside MT, Block SM. Reconstructing folding energy landscapes by single-molecule force spectroscopy. *Annu Rev Biophys.* 2014;43(1):19–39.
91. Bryngelson JD, Onuchic JN, Socci ND, Wolynes PG. Funnels, pathways, and the energy landscape of protein folding: a synthesis. *Proteins Struct Funct Genet.* 1995;21 (3):167–195.
92. Dudko OK, Graham TGW, Best RB. Locating the barrier for folding of single molecules under an external force. *Phys Rev Lett.* 2011;107(20):208301.
93. Du R, Pande VS, Grosberg AY, Tanaka T, Shakhnovich ES. On the transition coordinate for protein folding. *J Chem Phys.* 1998;108(1):334.
94. Chodera JD, Pande VS. Splitting probabilities as a test of reaction coordinate choice in single-molecule experiments. *Phys Rev Lett.* 2011;107(9):098102.
95. Neupane K, Manuel AP, Lambert J, Woodside MT. Transition-path probability as a test of reaction-coordinate quality reveals DNA hairpin folding is a one-dimensional diffusive process. *J Phys Chem Lett.* 2015;6(6):1005–1010.
96. Onoa B. Identifying kinetic barriers to mechanical unfolding of the *T. thermophila* ribozyme. *Science.* 2003;299(5614):1892–1895.
97. White KH, Orzechowski M, Fourmy D, Visscher K. Mechanical unfolding of the beet Western yellow virus −1 frameshift signal. *J Am Chem Soc.* 2011;133(25):9775–9782.
98. Hansen TM, Reihani SNS, Oddershede LB, Sørensen MA. Correlation between mechanical strength of messenger RNA pseudoknots and ribosomal frameshifting. *Proc Natl Acad Sci USA.* 2007;104(14):5830–5835.
99. Chen G, Chang K-Y, Chou M-Y, Bustamante C, Tinoco I. Triplex structures in an RNA pseudoknot enhance mechanical stability and increase efficiency of −1 ribosomal frameshifting. *Proc Natl Acad Sci USA.* 2009;106(31):12706–12711.
100. Green L, Kim C-H, Bustamante C, Tinoco I. Characterization of the mechanical unfolding of RNA pseudoknots. *J Mol Biol.* 2008;375(2):511–528.
101. Ritchie DB, Foster DAN, Woodside MT. Programmed −1 frameshifting efficiency correlates with RNA pseudoknot conformational plasticity, not resistance to mechanical unfolding. *Proc Natl Acad Sci USA.* 2012;109(40):16167–16172.
102. Ritchie DB, Soong J, Sikkema WKA, Woodside MT. Anti-frameshifting ligand reduces the conformational plasticity of the sars virus pseudoknot. *J Am Chem Soc.* 2014;136 (6):2196–2199.
103. de Messieres M, Chang J-C, Belew AT, Meskauskas A, Dinman JD, La Porta A. Single-molecule measurements of the CCR5 mRNA unfolding pathways. *Biophys J.* 2014;106 (1):244–252.
104. Mazauric M-H, Seol Y, Yoshizawa S, Visscher K, Fourmy D. Interaction of the HIV-1 frameshift signal with the ribosome. *Nucleic Acids Res.* 2009;37(22):7654–7664.
105. Mazauric M-H, Leroy J-L, Visscher K, Yoshizawa S, Fourmy D. Footprinting analysis of BWYV pseudoknot-ribosome complexes. *RNA.* 2009;15(9):1775–1786.
106. Yan S, Wen J-D, Bustamante C, Tinoco I. Ribosome excursions during mRNA translocation mediate broad branching of frameshift pathways. *Cell.* 2015;160(5):870–881.
107. Seol Y, Skinner G, Visscher K. Elastic properties of a single-stranded charged homopolymeric ribonucleotide. *Phys Rev Lett.* 2004;93(11):118102.
108. Vanzi F, Takagi Y, Shuman H, Cooperman BS, Goldman YE. Mechanical studies of single ribosome/mRNA complexes. *Biophys J.* 2005;89(3):1909–1919.
109. Seol Y, Skinner GM, Visscher K, Buhot A, Halperin A. Stretching of homopolymeric RNA reveals single-stranded helices and base-stacking. *Phys Rev Lett.* 2007;98(15): 158103.

110. Pisarev AV, Kolupaeva VG, Yusupov MM, Hellen CU, Pestova TV. Ribosomal position and contacts of mRNA in eukaryotic translation initiation complexes. *EMBO J.* 2008;27(11):1609–1621.
111. Brochard F, de Gennes PG. Dynamics of confined polymer chains. *J Chem Phys.* 1977;67 (1):52.
112. Odijk T. The statistics and dynamics of confined or entangled stiff polymers. *Macromolecules.* 1983;16(8):1340–1344.
113. Tree DR, Wang Y, Dorfman KD. Mobility of a semiflexible chain confined in a nanochannel. *Phys Rev Lett.* 2012;108(22):228105.
114. Wang J, Gao H. Stretching a stiff polymer in a tube. *J Mater Sci.* 2007;42(21):8838–8843.
115. Uemura S, Dorywalska M, Lee T-H, Kim HD, Puglisi JD, Chu S. Peptide bond formation destabilizes Shine–Dalgarno interaction on the ribosome. *Nature.* 2007;446(7134):454–457.
116. Wen J-D, Lancaster L, Hodges C, et al. Following translation by single ribosomes one codon at a time. *Nature.* 2008;452(7187):598–603.

CHAPTER THREE

Studies of RNA Sequence and Structure Using Nanopores

Robert Y. Henley*, Spencer Carson*, Meni Wanunu*,†,1
* Department of Physics, Northeastern University, Boston, Massachusetts, USA
† Department of Chemistry and Chemical Biology, Northeastern University, Boston, Massachusetts, USA
[1] Corresponding author: e-mail address: wanunu@neu.edu.

Contents

Abstract

Nanopores are powerful single-molecule sensors with nanometer scale dimensions suitable for detection, quantification, and characterization of nucleic acids and proteins. Beyond sequencing applications, both biological and solid-state nanopores hold great promise as tools for studying the biophysical properties of RNA. In

Progress in Molecular Biology and Translational Science, Volume 139
ISSN 1877-1173
http://dx.doi.org/10.1016/bs.pmbts.2015.10.020

this review, we highlight selected landmark nanopore studies with regards to RNA sequencing, microRNA detection, RNA/ligand interactions, and RNA structural/conformational analysis.

1. INTRODUCTION TO NANOPORES

Nanopores have garnered much attention for their ability to act as single molecule DNA sequencers when paired with a proper ratcheting enzyme and ionic current monitoring electronics. Despite their biological importance, much less work has been reported on the study of RNA molecules. Much like DNA, the direct sequencing of single RNA molecules is a topic of great importance. In addition, the secondary and tertiary structures formed by relatively short RNA fragments are also topics of equal biological importance. Nanopores are unique tools that have been shown to be extremely sensitive to variations in sequence as well as structure. In this chapter, we discuss the current state of RNA sequencing as well as structural interrogation using both solid-state and biological nanopores.

The principle of nanopore sensing is based on fast measurements of ion fluctuations through a nanoscale orifice. A cross-sectional view of a nanopore in an insulating membrane (not to scale) is shown in Fig. 1A. A concentrated electrolyte solution (>100 mM) is placed on both sides of the membrane such that the only electrical contact between the chambers is at the nanopore. An electrochemical bias applied across the membrane causes ion transport through the nanopore at rates that are determined by the nanopore dimensions, nanopore surface charge, applied voltage, ion mobility, and solution viscosity. In general for a thin pore of symmetric geometry, the current–voltage response is linear, as shown in Fig. 1A (bottom). Applying a constant DC bias results in a steady-state current that serves as a baseline measurement of the nanopore volume.

Transient entry and exit of biological macromolecules from the nanopore results in discrete disruptions in the steady-state baseline signal, as shown in Fig. 1B for the case of DNA molecules. Since nucleic acids are negatively charged the voltage applied to the analyte chamber (*cis* chamber) is more negative than the voltage applied to the *trans* chamber in order to electrophoretically drive the molecules through the nanopore. The transport kinetics of nucleic acids through nanopores is highly dependent on the nanopore geometry and other experimental conditions. While this topic has been

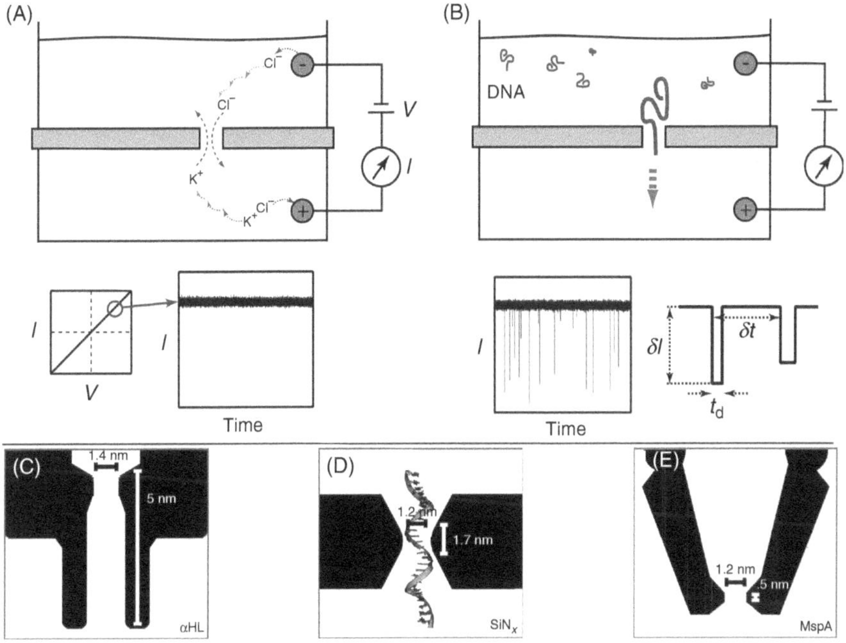

Figure 1 Schematic of the nanopore ionic current detection of biomolecules,[1] as well as illustrations of the most commonly used nanopores.[2] (A) A partition containing a nanoscale aperture (nanopore) separates two chambers containing electrolyte solution. When a voltage bias is applied across the partition, a steady ionic current through the nanopore is detected (images not to scale). (B) When DNA is added to the grounded (*cis*) chamber, it is electrophoretically driven through the nanopore to the positively charged (*trans*) chamber. The passage of each individual DNA molecule results in a transient decrease in the ionic current through the nanopore. (C–E) Schematics of α-hemolysin (αHL), silicon nitride (SiN), and *Mycobacterium smegmatis* porin A (MspA) nanopores (left to right) are shown with dimension labels (SiN pores have variable geometry). *Images obtained with permission from Venta et al.*[2]

studied exhaustively both theoretically and experimentally, it is beyond the scope of this chapter and the reader is referred to other reviews for a discussion on this topic.[1,3,4]

2. EARLY EXPERIMENTS TOWARD RNA SEQUENCING

2.1 Differentiation of RNA Homopolymers

It is of historical significance that the first experiment to report electrophoretically driven transport of biomolecules across a nanopore were done with homouridine RNA fragments (polyU).[5] In these pioneering experiments the investigators formed a lipid bilayer across a 100 μm diameter orifice in a polytetrafluoroethylene (PTFE) partition; the bilayer and partition are used to separate two chambers containing a buffered electrolyte solution. When αHL nanopores are added to one of the chambers (*cis*), one porin will spontaneously insert itself into the lipid bilayer. In the presence of a transmembrane potential, αHL insertion is observed as a discrete increase in ion current across the pore. Using αHL as a biomolecular sensor, researchers were able to detect the presence of RNA homopolymers (as well as DNA heteropolymers) as individual molecules translocated the nanopore. Kasianowicz et al. determined that the average time necessary for an RNA homopolymer to translocate the pore scales linearly with polymer length (ie, between 150 and 415 nt), thus demonstrating that nanopores can be used to measure the length of linear RNA fragments. The authors also hypothesized that this technique could be used to sequence DNA and RNA. While the former has been fully realized,[6,7] experimenters have been making steady progress toward the latter.

Since the first experiment, various research groups have made great strides in advancing the field of nanopore sensing by making favorable mutations to the αHL nanopore, as well introducing a number of new nanopore systems, both protein based and synthetic. The three types of nanopores we will discuss in this chapter can be seen in Fig. 1C–E, as adapted from Venta et al.,[2] with αHL and MspA nanopores being the biological examples and SiN nanopores being the synthetic example. We will not cover device fabrication in this chapter, but many reviews exist on this topic.[1,8]

2.2 Observation of polyU and polyC Segments Within Diblock Copolymers

A sequence specificity study by Akeson et al. demonstrated that the same αHL nanopores are able to distinguish between RNA bases based on the unique current blockade level each RNA base produces during translocation.[9] More specifically, their experiments revealed that RNA homopolymers of polyU, polyC, and polyA could be distinguished based upon the distinct current blockade of each molecule. The authors also showed that translocation of an RNA diblock copolymer containing repeats of A and C bases (ie, $A_{(30)}C_{(70)}$) produces bilevel events that are in good agreement with the current levels seen by polyA and polyC homopolymers. Interestingly, the authors noted that 90% of events produced blockades that correspond to 3′ entry of the polymer. The nature of this preferential orientation was later expanded upon by Butler et al. who observed that both C_{50} homopolymers as well as $A_{25}C_{50}$ coblock polymers favor entry from their polyC 3′ ends.[10]

2.3 DNA Sequencing Using an MspA Pore as a Reader and DNA Polymerase as a Molecular Ratchet

Over the next few years the promise of nanopores as a viable next-generation sequencing platform spurred development forward. The two major obstacles toward nanopore-based DNA sequencing were verification of true single base specificity and control over DNA transport during translocation. Through site-directed mutagenesis researchers were able to show that mutant αHL (E11N/K147N/M113Y)[11,12] as well as mutant MspA nanopores[13] are capable of individual DNA base recognition. Soon after, the Akeson and Gundlach groups slowed down DNA translocation to measurable rates (ie, milliseconds per nucleotide) by enzymatically ratcheting DNA base by base through a nanopore.[6,7] Further information on the progress toward nanopore-based DNA sequencing can be found in several reviews.[1,3,4]

2.4 Discriminating Single Bases Using Immobilized RNA Strands

Ayub and Bayley later demonstrated that the αHL pore is sensitive to single base substitutions in RNA fragments.[14] In this study the authors compared the wild-type (WT) αHL to the two αHL mutants that had previously shown sensitivity to base recognition. Here the authors used the same method that was previously employed to show that αHL and MspA nanopores are able to resolve all four DNA bases.[12,13] For this experiment four oligonucleotides were synthesized, each with a different ribobase at the ninth nucleotide position (counted from the 3′ end) within an otherwise polyU sequence, which was chosen to eliminate formation of secondary structures. Each oligo was produced with a biotinylated 3′ end that was then conjugated to a streptavidin molecule. The ninth nucleotide position was specifically chosen as the site of variation because once a streptavidin–RNA complex is drawn into the pore (Fig. 2A), the ninth nucleotide will sit precisely at the central constriction of the αHL nanopore. Previous experiments have shown that the WT αHL pore contains four constrictions or recognition sites, but the central constriction (referred to as R1) appears to be the most sensitive of the four sites. As such the authors measured the ionic current through the pore when each of the four ribonucleobases was immobilized at the R1 site. While the WT and mutant NN pore displayed some level of base specificity, they were unable to completely resolve each base. In contrast, the NNY

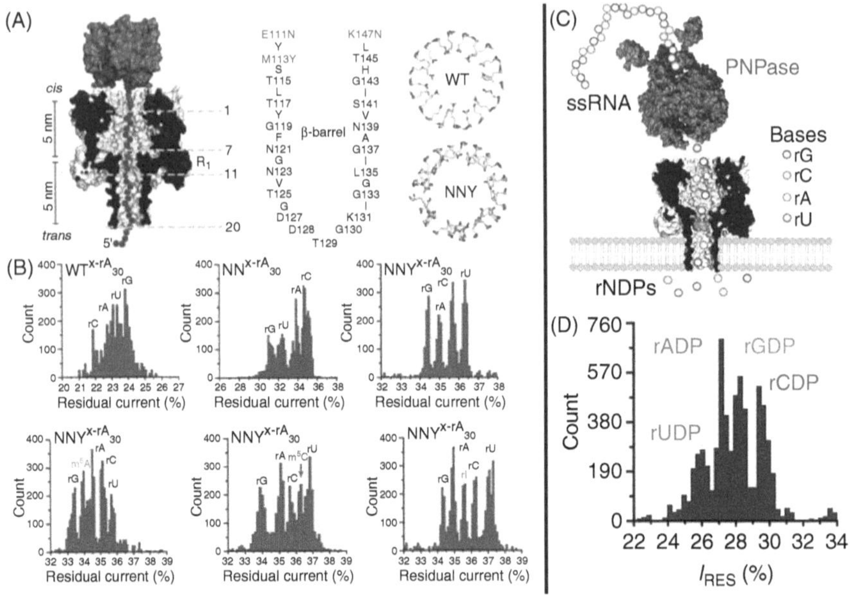

Figure 2 The road toward direct RNA nanopore sequencing. (A) An illustration of the streptavidin immobilized RNA strand used by Ayub et al.[14] The four proposed constrictions are labeled, and diagrams of the WT and NNY mutant αHL pores are shown. (B) Histograms of the residual ionic current while an RNA is immobilized within the nanopore are displayed. The top row shows the superior base discrimination of the NNY mutant when compared to WT and NN pores. The bottom row demonstrates the additional epigenetic sensing ability of the NNY mutant. (C) A diagram of the proposed exonuclease-assisted sequencing method.[15] (D) Current histogram comparing the residual currents caused by binding of rADP, rGDP, rCDP, and rUDP to the molecular adapter inside the αHL β-barrel. *Images obtained with permission from Ayub et al.*[14,15]

mutant produced substantial separation in the current levels observed for each base. The authors were also able to produce similar results for the N^6-methyladenosine (m^6A), 5-methylcytosine (m^5C), and inosine (I) post-transcriptional modifications. By repeating the previous experiment with these modified bases at the ninth nucleotide position, they observed that each base has a current blockade level that is distinct from each of the four canonical RNA bases (Fig. 2B).

2.5 Progress Toward Enzymatic Control of RNA Transport

While φ29 polymerase has been shown to be a suitable enzyme for controlling DNA motion by ratcheting DNA through nanopores, an enzyme for the ratcheting of RNA through nanopores is yet to be presented. Before the demonstration of nanopore sequencing via φ29 polymerase, researchers proposed to use an exonuclease-based method for nanopore sequencing. In this method, a DNA/exonuclease complex is electrophoretically captured by a nanopore where the exonuclease will sequentially cleave the DNA bases one at a time (Fig. 2C). As the bases are cleaved, the applied electric field pulls the bases to the pore where they encounter a molecular adapter that temporarily traps the base, and identifies it according to its unique current blockade.[16] This method was revisited for the purpose of RNA sequencing and researchers showed that, as with DNA, this method can be used to accurately discriminate between individual RNA bases (Fig. 2D).[15] Unfortunately, as with the exonuclease-based DNA sequencing method, significant problems persist with this method pertaining to the efficiency of capturing cleaved nucleotides.[17] Once nucleotides are cleaved by the exonuclease, the electrophoretic force biases their motion toward the pore, but Brownian forces hamper this directionality. As a result of this diffusive effect, some bases will diffuse into the bulk solution and not be read by the nanopore, or be captured after diffusing out of the pore. This results in bases being read out of sequence, as well as insertion and deletion errors. Several improvements to this method have been proposed to circumvent the diffusion problem,[18] but to date this remains an open problem.

2.6 Summary

The experiments described earlier represent an important step toward nanopore-based RNA sequencing. A single-molecule method for direct RNA sequencing that is able to resolve RNA modifications would be a valuable tool for medical diagnoses as well as academic studies of the regulatory roles played by these modifications. However, it remains to be seen whether these modifications can be resolved from one another in the context of a heavily modified RNA sequence, if these developments can be translated into a nanopore device with a single recognition site capable of sequencing or if there is an enzyme that is capable of long-read length ratcheting of RNA molecules. Though major roadblocks for this method persist, realization of such a device would certainly change the landscape of transcriptomics.

3. POINT-OF-CARE APPLICATIONS

Though nanopores have received a great deal of attention for their promise as tools for sequencing nucleic acids, they have also shown to be extremely effective platforms for potential single-molecule point-of-care (POC) systems. The nanopore's extreme sensitivity to small differences in the charge and size of passing molecules makes them excellent systems for the detection of nucleic acid probes, RNA–ligand interactions, antibiotic complexes, etc. In addition the single-molecule nature of these systems allows them to perform rapid analysis without the need for error prone amplification steps or time consuming and costly labeling procedures.

3.1 MicroRNA Detection and Quantification

3.1.1 A Solid-State Nanopore System for Rapid Detection of Target microRNA Expression

One promising application for nanopores is the rapid detection of microRNA (miRNA) molecules. Several methods have been developed for the detection and quantification of miRNA molecules and many more are currently in development.[19] The most conventional methods involve probe labeling using radioisotopes and/or some form of PCR amplification. Though these methods are effective, the use of radioactive labeling is hazardous and often times inconvenient, and amplification can be error prone and time consuming.

The Drndić Lab has demonstrated a promising technique for the analysis of miRNAs using solid-state nanopores,[20] which takes advantage of the sensitivity of the nanopore platform to Angstrom differences in molecule size. The authors showed that by fabricating 3-nm diameter nanopores in thinned solid-state membranes (ie, 6-nm effective thickness) they can distinguish between dsDNA, dsRNA, and transfer RNA (tRNA) molecules, and previous papers have demonstrated the ability of solid-state nanopores to distinguish between single-stranded and double-stranded nucleic acids.[21] This is accomplished by monitoring the ionic current through the pore for spontaneous reductions, each of which are representative of the diameter of the translocating molecule. By comparing the current blockades produced by translocation of an unknown sample with those produced by the translocation of known single-stranded and double-stranded samples, it becomes possible to identify the unknown sample based on its diameter. This same feature of nanopores is exploited to identify target miRNA sequences within total RNA extracts.

In this method, RNA extracted from tissue cells is incubated with probe RNA molecules whose sequence is complimentary to that of the target molecule and therefore binds to any target miRNA molecule present.

This sample is then incubated with magnetic beads that are functionalized with the p19 protein from the Carnation Italian ringspot virus. This protein has been shown to selectively bind to dsRNA in a size-dependent, but sequence-independent manner, specifically binding to 21–23 bp dsRNA. Accordingly, the p19 protein will bind specifically to any probe/miRNA duplex molecules present in the sample. The sample is then washed to remove the remaining RNA molecules that are not bound to the magnetic beads and finally the desired probe/miRNA duplex is eluted from the protein (see Fig. 3A for a diagram of this entire process).

Once the probe/miRNA duplex has been isolated, these ultrathin nanopore platforms can easily distinguish any dsRNA molecules present, thereby verifying the presence of the target miRNA. Example nanopore traces of this sample (RL), a positive control (PC), and negative controls (NC) are shown in Fig. 3B. In order to quantify the amount of target miRNA present in a sample, a calibration curve is constructed. The calibration curve is populated by measuring the rate of translocation events from control dsRNA samples with varying concentrations. From this calibration curve the authors concluded that there is a linear relationship between the rate of translocation events and the concentration of molecules in the sample. Finally, when an unknown sample is tested its event rate can easily be linked to a corresponding target miRNA concentration, as shown in Fig. 3C. The authors confirm that this method can quantify a 1 fM sample in ~4 min, making it a promising platform for the rapid detection of biologically relevant levels of miRNA.

3.1.2 Detecting Lung Cancer-Related miRNAs and Single Nucleotide Polymorphisms Using a Biological Nanopore System

The Gu group have presented a similar probe-based approach to quantification of miRNAs using an αHL nanopore.[22] This method does not use a simple threshold technique for detecting probe bound molecules, as the αHL nanopore constriction is too narrow to allow the passage of dsRNA. Instead, this method relies upon the identification of what the authors call a "signature signal." To produce this signature signal, the authors use a probe molecule that is similar to the one used by the Drndić Lab in that it is designed to target a specific miRNA sequence by Watson–Crick base pairing in its central region, but is extended from both its 3′ and 5′ ends with a polydC$_{30}$ tag. The capture of this probe/miRNA complex produces a distinctive signal, which is shown in detail in Fig. 3E. As the complex enters the vestibule of the pore the extended end of the probe molecule threads

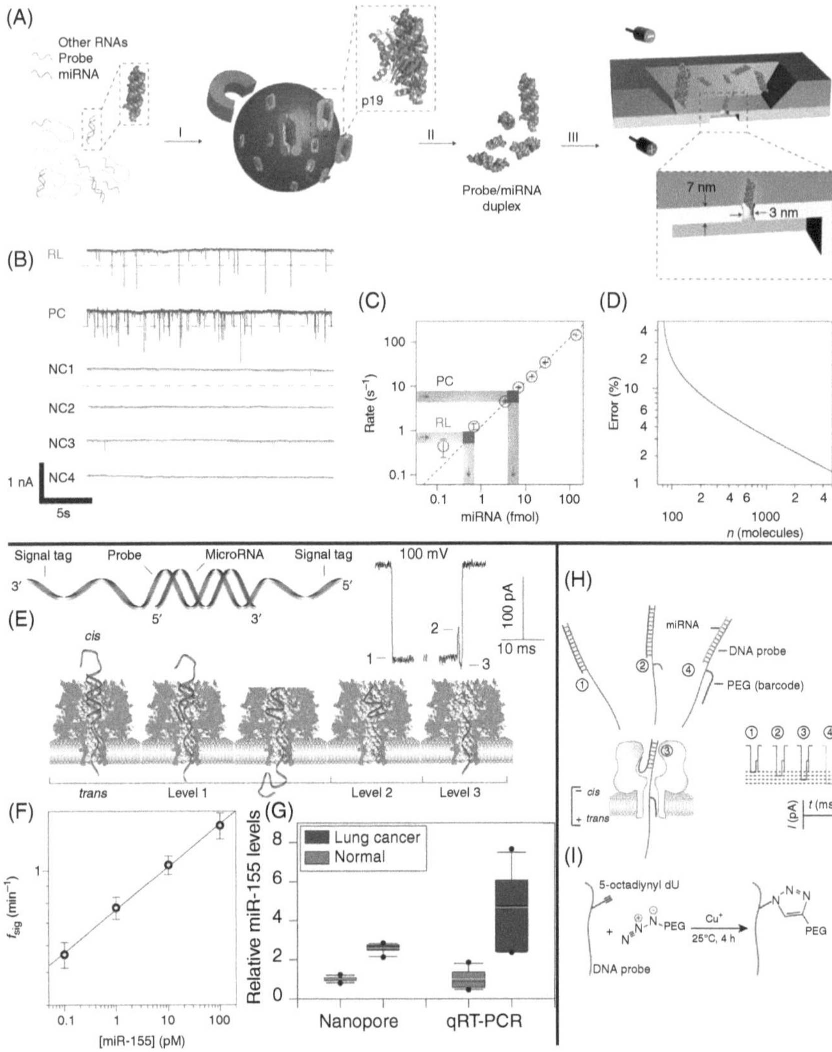

Figure 3 Nanopore-based miRNA detection schemes. (A) A cartoon of the miRNA detection scheme used by the Drndić Lab.[20] Total RNA extract is incubated with probe molecules complimentary to a target miRNA sequence. This mixture is then incubated with magnetic beads that are functionalized with the p19 protein, which selectively binds to 21–23 bp dsRNA molecules. A washing step then isolates the target miRNA/probe complex, which can be readily detected using a thin SiN nanopore. (B) A sample of the ionic current traces resulting from translocations of the isolated miRNA/probe complexes. RL is the trace resulting from enrichment of miR122a, PC is a positive control, and NC refers to various negative control experiments. (C) A calibration curve is constructed by plotting the event rates as a function of miRNA concentration. The event rates of the PC and RL experiments are shown as indicated and the miRNA concentration

through the β-barrel of the pore (Level 1), here the miRNA is sterically required to "unzip" from the probe molecule as the probe translocates, the current momentarily increases as the probe exits the β-barrel region (Level 2), then quickly reduces to a similar level as the miRNA enters the β-barrel (Level 3), and finally returns to the open pore level upon translocation of the miRNA. Counting the rate of these signature events, and using a similar calibration process as described by the Drndić Lab enables the authors to quantify miRNA concentrations as low as 100 fM (Fig. 3F).

Notably, this platform is capable of detecting single nucleotide mismatches between the miRNA and probe sequence, corresponding to miRNA single nucleotide polymorphisms (SNPs). A fully complementary probe molecule will bind more tightly to the target miRNA and thus take longer to unzip than the molecule containing an SNP. The authors demonstrated that by first calibrating the system with a fully complimentary prove/miRNA complex, they are able to detect three different lung cancer-related SNPs, as each SNP results in a distinct decrease in the unzipping time.

The authors used this technique to detect the presence of miR-155, an miRNA that has elevated levels in lung cancer patients. The results obtained from this method matched the diagnosis of those obtained from qRT-PCR, but with a higher degree of precision (Fig. 3G).

3.1.3 Multiplexing microRNA Detection Using Probe Specific Tags

The Gu Lab recently published a paper representing a significant advance in the development of miRNA detection technology.[23] The authors demonstrate a method for overcoming the problem of simultaneous detection of several miRNA targets by conjugating PEG tags of varying lengths to the probe molecules through copper(I)-catalyzed click chemistry, with each PEG length corresponding to a different miRNA (Fig. 3H and I). This method relies on the ability of PEG to modulate the ionic current, and hence producing characteristic current levels depending upon the size of

◀ is inferred from the calibration curve. (D) The error in concentration as predicted from the nanopore experiments, as a function of number of translocation events. (E) A diagram of the unzipping method used by the Gu Lab.[22] A probe/target miRNA complex is shown along with the translocation process of this molecule and its resulting ionic current trace. (F) A calibration curve constructed by plotting event rate versus miRNA concentration is shown. (G) The miRNA levels as detected by the nanopore, as well as qRT-PCR are compared. (H) A schematic of the revised Gu Lab strategy,[23] employing four different polyethylene glycol (PEG) tags. (I) A diagram of the click chemistry tagging process. *Images obtained with permission from Wanunu et al.*[20,22]

the translocating PEG molecule.[24,25] As probe/miRNA complexes translocate the pore, each produces a "signature signal," similar to that described earlier, but with the addition of a variable-length PEG tag discrimination of several miRNA targets is possible. This tagging system is used to simultaneously detect the concentrations of four different miRNAs down to 10 pM (though, the authors suggest a limit of 100 fM).

Though these experiments represent significant progress toward a mature nanopore miRNA detector, a number of problems must still be addressed. The Gu Lab has shown simultaneous detection of four different miRNAs, but an ideal miRNA detection platform would be able to detect hundreds or thousands of miRNA in parallel. This need dictates that any mature system should incorporate a large array of nanopores, though the strategy proposed by the Gu Lab can simplify this engineering task by decreasing the size of the array by at least a factor of four. The optical detection scheme developed by the Wallace and Bayley labs (but not covered in this chapter) may be able to address this issue.[26] Additionally, other proposed miRNA detection schemes have demonstrated attomolar detection limits.[27] While such a low detection limit may be unnecessary, demonstration of a similar sensitivity would prove useful for detection of rare miRNAs or early stage diagnoses. Many techniques have been developed to push nanopore detection limits,[28–30] but none have been demonstrated in conjunction with miRNA quantification to date.

3.2 RNA/Ligand Interactions and Nanopore-Based Drug Screening

3.2.1 Antibiotic Drug Screening Through Detection of RNA/Drug Binding

Another potential POC application for the nanopore platform is as a monitor for RNA/ligand interactions. The Drndić Lab have presented a proof-of-principle experiment showing that the previously described thin solid-state nanopore system can be used as a platform for high-throughput drug screening assays, particularly for RNA-mediated diseases.[31] To perform this first demonstration, the authors tested the ability of nanopores to detect the binding of several aminoglycoside antibiotics. Aminoglycosides target the prokaryotic ribosomal RNA decoding site or A site and are of clinical importance in treating bacterial infections. For these nanopore experiments, a truncated RNA construct containing the A site was used as the RNA target. Previous experiments confirm that this truncated RNA construct mimics the function of the intact A site and it interacts with ligands and antibiotics in a manner that correlates with normal subunit function.[32–34]

Neomycin, paromomycin, and kanamycin are three such antibiotic agents that have been shown to bind to this truncated construct and are clinically effective in disrupting protein synthesis. To first characterize the sensitivity of the nanopore system to drug-binding events, bare RNA was first translocated through a 3-nm diameter SiN pore ($N = 8552$); this process was then repeated in the presence of 70 μM paromomycin ($N = 5286$) and 130 μM paromomycin ($N = 6519$) (Fig. 4A). Histograms were then constructed for

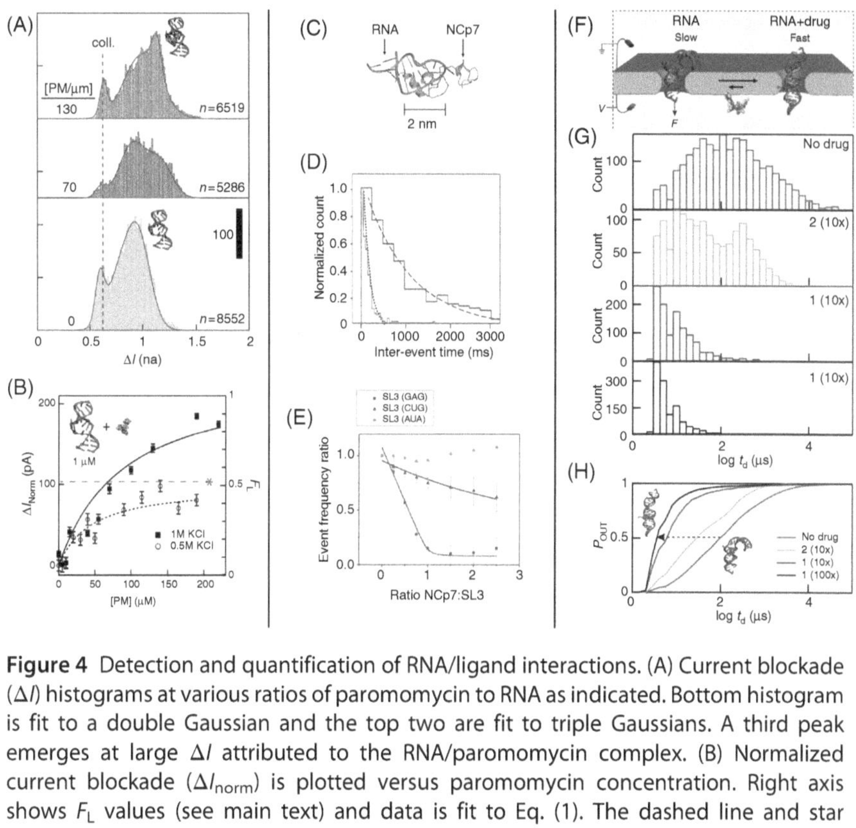

Figure 4 Detection and quantification of RNA/ligand interactions. (A) Current blockade (ΔI) histograms at various ratios of paromomycin to RNA as indicated. Bottom histogram is fit to a double Gaussian and the top two are fit to triple Gaussians. A third peak emerges at large ΔI attributed to the RNA/paromomycin complex. (B) Normalized current blockade (ΔI_{norm}) is plotted versus paromomycin concentration. Right axis shows F_L values (see main text) and data is fit to Eq. (1). The dashed line and star indicate the saturated values at 0.5 M KCl. (C) Diagram of the RNA/NCp7 complex used by the Movileanu group.[35] (D) Histograms of the interevent times for RNA only [dotted (purple in the web version) curve] and with added NCp7 [dashed (red in the web version) curve]. (E) Event frequencies normalized to RNA; only event rate are plotted against the ratios of NCp7/RNA constructs. (F) A cartoon diagram of the hepatitis C virus (HCV) conformational analysis system used by the Wanunu Lab.[36] (G) Histograms showing the log of the dwell time (t_d) for various drug concentrations as indicated. (H) Integrated normalized dwell-time distributions are plotted for various drug concentrations. *Images obtained with permissions from Wanunu et al.,[31] Niedzwiecki et al.,[35] and Shasha et al.[36]*

the mean value of the current blockade from each event. A persistent low blockade peak can be observed at a value of ~0.6 nA for all samples, which the authors attributed to failed attempts by the molecules to translocate the pore, or "collisions." Each sample also displayed a population near ~0.9 nA, representing bare RNA translocations. While the control sample containing no paromomycin fits very well to a double-Gaussian distribution, a third peak seems to develop with increasing levels of paromomycin near ~1.15 nA, which is most likely due to binding of paromomycin to the A-site molecule. As the complexed molecule passes through the pore, it blocks more of the ionic current than that of the A-site molecule alone, mainly due to an increase in the excluded volume of ions.

To quantify the binding affinity of paromomycin to the A site and construct a binding isotherm, it is necessary to know the fraction of drug-bound RNA molecules. Assuming a simple 1:1 binding model (ie, Drug + RNA $\rightleftharpoons$ Drug:RNA) and equal capture rates for the RNA and RNA/drug complex, a simple relationship for the dissociation rates can be inferred. For such a system, the dissociation constant is K_d = [Drug][RNA]/[Drug:RNA]. With the dissociation constant we can calculate the fraction of RNA molecules that form a complex with the drug as a function of drug concentration (F_C).

$$F_C = \frac{[\mathrm{Drug}]}{[\mathrm{Drug}] + K_d} \tag{1}$$

As can be seen in Fig. 4A, the bare RNA and RNA/drug complex have largely overlapping current blockade distributions; as a result it is not possible to simply count the fraction of bound molecules. Although, the authors find that they can effectively quantify the fraction of bound molecules by monitoring the mean of the current blockade distribution and plotting its deviation from the mean current blockade of the bare RNA molecule (ΔI_{norm}) as a function of drug concentration. To estimate the fraction of molecules bound, first ΔI_{norm} is measured with drug concentration in excess to RNA concentration to obtain the current blockade for a completely saturated system (ΔI_{max}). The fraction of molecules bound is then found by normalizing ΔI_{norm} to this saturated value ($F_L = \Delta I_{\mathrm{norm}}/\Delta I_{\mathrm{max}}$). By fitting the ΔI_{norm} versus drug concentration data to Eq. (1), as displayed in Fig. 4B, it is possible to extract K_d for paromomycin. Using this analytical method, the authors estimated a K_d value of 90 ± 34 μM in the presence of 1 M KCl and 44 ± 17 μM in the presence of 0.5 M KCl. The increase in affinity of aminoglycosides (such as paromomycin) to the A-site molecule in the presence of reduced salt concentrations is a well-documented effect resulting from the decreased screening of electrostatic interactions.[34] This same method was also used to extract K_d values for two additional A-site targeting antibiotic molecules, neomycin and kanamycin. A K_d value of 14.4 ± 3.3 μM is reported for neomycin at 1 M KCl, while the observed K_d value of kanamycin was too low to be observed at this salt concentration. These measurements are consistent with previous observations made using an independent FRET-based assay, showing that at 0.5 M NaCl, neomycin has a much higher affinity to the A-site construct than paromomycin, which has a much higher affinity than kanamycin.[37]

In the system presented here, a statistically significant amount of data (~500 events) is collected in under 1 min using ~1 pmol of RNA, demonstrating that nanopores may be used as an effective tool for rapid drug screening assays.

3.2.2 A Nanopore Assay for the Detection of an RNA-Binding HIV Biomarker

The Movileanu group has shown that nanopores can be used to monitor RNA–ligand interactions even when the size of the ligand does not permit transport of the RNA/ligand complex through the nanopore.[35] In this work, SiN nanopores are used to detect binding of the human immunodeficiency virus 1 (HIV-1) nucleocapsid protein p7 (NCp7)[38] to RNA aptamers. NCp7 is a protein that is involved in the packaging of the HIV-1 viral genome and in functions critical to viral replication. This protein contains two zinc finger domains that recognize the packaging domains of genomic RNA by preferentially binding to exposed guanosines in RNA stem loops (Fig. 4C). To test the nanopore's ability to detect the binding of NCp7, translocation experiments were conducted with three 20-nucleotide RNA aptamers with various degrees of binding affinity. These aptamers were derived from a region of the retroviral ψ-packaging element of the HIV-1 genome known as stem-loop 3 (SL3). The aptamers derived from the SL3 region are known to bind the NCp7 protein with high affinity. The aptamers designed for this experiment form stem-loop structures, but each varies in the number of guanosine bases in the stem loop, thus varying its binding affinity to NCp7.

Two methods for calculating binding affinities are described by the authors: one using nanopores that are small enough to deny the passage of the RNA/protein complex, but large enough to translocate the bare RNA aptamer (diameter < 6 nm) and another method using large nanopores that are able to translocate the complex (diameter of 7–15 nm). First using small nanopores, translocation data was collected for the standard SL3 aptamer (GAG) at a concentration of 500 nM and then with the NCp7 protein added to the system at a concentration of 500 nM. Since the protein/RNA complex is too large to translocate the pore, a decrease in the translocation event rate was observed upon addition of NCp7 since only the fraction of molecules not bound by NCp7 produce translocation events. Creating a histogram of the interevent time and fitting the data to single exponential distributions (Fig. 4D) show that the interevent time increases from 149 ± 12 to 1030 ± 60 ms upon addition of NCp7, which confirms the binding of NCp7 to the GAG aptamer. This technique is then used to extract the binding affinities of NCp7 to each of the RNA aptamers by constructing titration curves of the interevent rates for each aptamer, which are presented in Fig. 4E. Though the binding affinities calculated using nanopores do not agree quantitatively with those found using FRET techniques,[39] they do agree qualitatively with FRET results, as NCp7 was observed to bind more tightly to aptamers with greater guanosine concentrations in their stem-loop regions (ie, GAG > CUG > AUA).

The authors also provide evidence that this system may be used to test HIV-1 drug candidates that target the NCp7 protein. By introducing an excess of *N*-ethylmaleimide (a compound that is known to inhibit the binding of NCp7) to the system, they are able to restore the event frequency to a level that is comparable to the rate observed before the addition of NCp7.

These results show that using nanopores small enough to prohibit the translocation of the target complex, one can extract binding affinity information solely by measuring the interevent times. In addition, the authors utilized nanopores that are large enough to translocate the RNA/protein complex (diameter $\sim 7-15$ nm) and observed that the addition of NCp7 results in an increase in events that block significantly more current than that observed for the RNA molecule alone (data not shown here). This result is very similar to that described earlier by the Drndić Lab,[31] except it does not find equal capture rates for the RNA and RNA/protein complex. As a result, a more complex equation is derived to fit the titration curve and the authors extracted a K_d value for NCp7 affinity to the GAG aptamer using the same experimental protocol as described by the Drndić Lab.

3.2.3 Screening for Conformational Changes Induced by RNA Drug Targets

The techniques presented by the Drndić and Movileanu labs represent two effective methods for detecting the binding affinities of ligands that add a considerable size to their target molecule. However, the Wanunu Lab has recently shown that it is still possible to detect RNA/ligand interactions, even when the binding molecule does not add considerable size, provided that there is some conformational change induced in the target molecule by the binding of the ligand.[36] The internal ribosome entry site (IRES) motif of the hepatitis C virus (HCV) was employed to demonstrate the sensitivity of nanopores to conformational changes in a target biomolecule. The HCV IRES domain is a highly conserved motif with a structure that promotes high affinity interactions with the 40S subunit of its host ribosome.[40] These interactions allow the HCV to hijack the host translational machinery and use it to translate the viral genome instead. A class of benzimidazole-based HCV inhibitors have been shown to disrupt HCV RNA translation.[41] These inhibitors bind to the HCV IRES subdomain IIa and induce a conformational change that results in widening of the RNA interhelical angle.[42] This conformational change promotes undocking of the IIa subdomain from the ribosome, thus preventing HCV translation. FRET has been an effective tool for the screening of potential inhibitors,[43] as it can be used to report on conformational changes in the nanometer range, but it requires tedious and often expensive labeling steps.

The Wanunu Lab has shown that nanopores can be used to detect the conformational changes that are induced by these benzimidazole derivatives in a rapid and label-free method, eliminating some of the drawbacks of FRET-based methods. As seen in Fig. 4F, translocation of the HCV subdomain IIa RNA construct is sterically hindered when using a ~3 nm diameter nanopore, while passage of the construct after it adopts its drug-induced linear conformation is largely unhindered. To sense this conformational shift, nanopore experiments were conducted with the RNA construct in its natively bent conformation, as well in the presence of two benzimidazole derived drugs that bind with different affinities (**1**: $K_d = 3.4\ \mu M$, **2**: $K_d = 64\ \mu M$, measured by an independent FRET assay[42,44]). Dwell-time histograms in the presence of different drug concentrations are shown in Fig. 4G. The peak in the dwell-time distribution for the RNA molecule alone appears to be around 100 μs. Upon addition of drug **2** (10:1 ratio of drug: RNA), there is a slight shift in the distribution, with a second population that develops near 10 μs. Addition of drug **1** (also in a 10:1 ratio) almost completely eliminates the longer timescale population and when the concentration is increased to 100:1 for drug **1** all events display timescales below 100 μs. To visualize this in a more quantitative fashion, the integrated normalized dwell-time distributions are plotted in Fig. 4H. The relative inflection of these curves match with what would be theoretically expected based on the binding affinity of the two drug molecules. Since drug **1** has a much lower dissociation constant than drug **2**, it binds to the RNA construct more tightly, resulting in more RNA constructs in the linearized conformation that translocate the nanopore faster. Though this study does not contain enough data points to extract accurate measurements of the dissociation constants, it does show that solid-state nanopores of a proper geometry can be used for rapid, label-free drug-screening assays to detect conformational shifts in target molecules.

3.3 Summary

The collection of papers discussed here demonstrate the ability of appropriately sized nanopores to sense and quantify RNA/ligand interactions. This platform offers a variety of sensing modalities, as each parameter (ie, current blockade, interevent time, or dwell time) has been shown to shift in response to different ligand-induced changes in the target molecule (ie, internal structure occupancy,[31] increased cross-sectional area,[35] or conformational shifts[36]). In each of the cases presented the nanopore provides rapid sensing of RNA/ligand interactions without the need for application-specific labeling techniques, making it a valuable tool for drug-screening assays as well as studies of molecule interactions.

4. RNA STRUCTURAL ANALYSIS

Though Shasha et al. were the first to demonstrate the sensitivity of nanopores to ligand-induced conformational changes, there are several other examples of nanopore-based RNA structural analysis. Borrowing from studies of nanopore DNA unzipping,[45,46] theoretical studies have shown that nanopores may be able to infer details about the sequence of an RNA molecule by sequentially unzipping and "ironing out" the secondary structure.[47,48] In addition, it has already been shown that solid-state nanopores are able to discriminate between ssDNA and dsDNA,[21] as well as dsDNA and dsRNA[20] based on their differing helical diameters. Here we will highlight a few of these important nanopore-based RNA structural analysis papers.

4.1 Observation of Helix-Coil Conformational Fluctuations

The first nanopore unzipping study to focus on RNA structures was conducted by Lin et al.,[49] in which they aimed to observe the helix-coil transition of a polyA RNA molecule on the single-molecule level. Temperature-jump experiments have been able to measure the submicrosecond conformational fluctuations between helix and coil conformations of polyA molecules at low-ionic strengths (50 mM salt).[50] To observe this conformational shift and accurately measure its timescale using an αHL nanopore, it is beneficial to increase the time that the molecule resides in the pore (polyA has been observed to translocate αHL at a rate of ~3 μs/nt with an applied bias of 120 mV at room temperature).[9] To increase the dwell time, the authors employed the DNA unzipping strategy as used in earlier nanopore studies.[45,46] By designing duplex nucleic acid molecules with single stranded overhangs (similar to those later used by the Gu Lab for miRNA detection[22,23]), the time needed for a molecule to translocate is significantly prolonged. The researchers designed three partial duplex molecules, each of which contains an identical 10 bp duplex region, but differs in its overhang region: (1) dA_{50}, 50 nt polydA overhang; (2) C_{25}, 25 nt polyC overhang; and (3) A_{25}, 25 nt polyA overhang. These constructs extend the dwell time of each molecule because the duplex region is too large to thread into the β-barrel and must completely unzip before the single-stranded segment can fully translocate the pore (Fig. 5A).

Employing this unzipping strategy, the authors observed dwell times of ~5 ms for the dA_{50} and C_{25} constructs. Interestingly, the A_{25} construct displays dwell times that are nearly an order of magnitude longer than dA_{50} and C_{25}. In addition to these longer timescales, large fluctuations can be observed in the ionic current when the A_{25} molecule occupies the nanopore cavity. To visualize these fluctuations, all-point current histograms are plotted from several representative translocation events of each molecule (Fig. 5B). As previously described, the authors observe that each of the polynucleotides exhibit a unique current blockade level, which can be seen in the all-point histograms (dA_{50}, C_{25}, and A_{25} display large peaks at 16.8, 19.6, and 21 pA, respectively). In addition to these characteristic peaks, a second peak can be seen in the all-point histogram for A_{25}. In the publication, it is theorized that the ionic current fluctuations observed must be the result of the translocating molecule oscillating between helix and coil conformational states, with the lower peak observed at 8.4 pA being the result of the more dense helix structure. This hypothesis also explains the increase in dwell time for the A_{25} construct, since the helix form may slow down the translocation process due to increased interaction with the pore walls, as well as steric considerations.

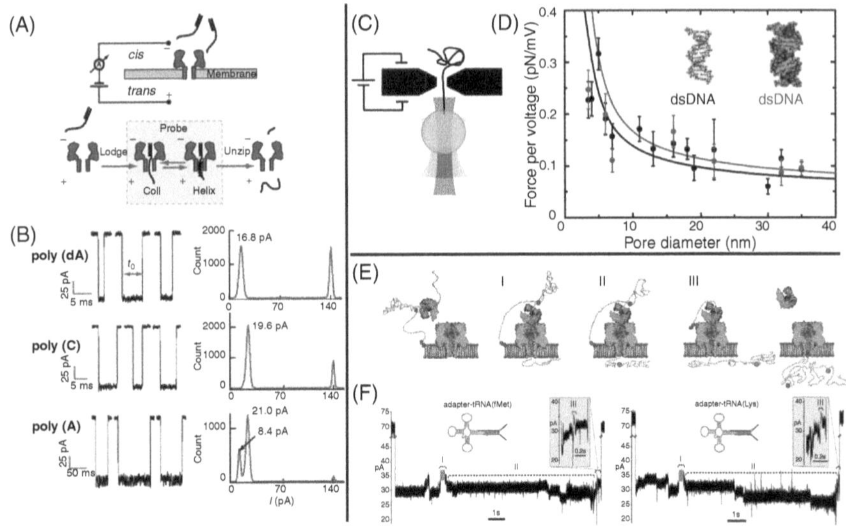

Figure 5 RNA structural analysis using nanopores. (A) A schematic of the unzipping technique used to observe the helix-coil transition.[49] (B) Characteristic sample events are shown for each of the molecules studied, with each accompanied by an all-point histogram of the current data. (C) A cartoon of the hybrid optical tweezers/solid-state nanopore setup used by the Cees Dekker Lab.[51] (D) Force per voltage as a function of nanopore diameter is plotted for both dsDNA (black) and dsRNA [gray (blue in the web version)]. (E) A schematic for the tRNA unfolding and transport process used by Smith et al.[52] (F) Characteristic ionic current traces observed during the capture, unfolding, and translocation of $tRNA^{fMet}$ and $tRNA^{Lys}$. *Images obtained with permission from Van Den Hout et al.*[51]

To provide additional evidence that the ionic current fluctuations observed were in fact due to the helix-coil transition, the authors next constructed two additional RNA molecules: (1) C_5A_{35} and (2) $C_{15}A_{25}$, where both possess the same 10 bp duplex region as the previously discussed molecules. Lin et al. hypothesized that the addition of a spacer larger than 12 nt should eliminate the two-state fluctuations in the ionic current, as the β-barrel region of the αHL nanopore is only ~5 nm long. This implies that the observed fluctuations should decrease as the length of the spacer increases as more of the helix-forming region is located outside the barrel region during the unzipping process. Indeed this theory was confirmed with all-point histograms showing slightly less two-state behavior for C_5A_{35} when compared to A_{25}, and $C_{15}A_{25}$ being virtually indistinguishable from C_{25} (data not shown).

Using a simple threshold analysis, the lifetimes of each of the two states was extracted from >2000 transitions. Fitting the lifetime data to exponential distributions, the mean lifetimes were found to be: $\tau_{helix} = 1.0$ ms and $\tau_{coil} = 3.1$ ms. Interestingly, these timescales are nearly three orders of magnitude slower than those observed by laser temperature-jump experiments.[50] The authors theorize that this increase in lifetime is due to stabilizing effects that arise from the confinement of the molecule in the nanopore. Finally, by conducting a temperature-dependent study, enthalpic barriers and free energy changes are calculated for the helix-coil transition as well as pore/RNA interaction energies for the two states. Measurements such as these may be extremely beneficial in understanding the effects of confinement of nucleic acids and their conformational states.

4.2 A Hybrid Optical-Tweezer/Nanopore System for Studying Confinement Effects on RNA

Optical tweezers have been a valuable tool in the study of secondary and tertiary structures formed by nucleic acids.[53] By optically trapping one end of a nucleic acid structure and attaching the other to some manipulator (micropipette, glass cover slip, etc.), one can stretch the captured molecule and monitor its response to the applied force (Fig. 5C). Alternatively, it is possible to optically trap both ends of the nucleic acid structure and measure changes in the positions of the optically trapped ends in response to some outside force (eg, proteins, ligands, etc.). These systems have been used to study RNA secondary and tertiary structures through force-induced melting,[54] as well as real-time monitoring of RNA folding events.[55] Optical tweezers have been combined with a number of different tools to provide an additional method for manipulating the molecule in question or to establish multiple, concurrent measurement systems. For a more in-depth account of RNA force measurements using optical tweezers, see the chapter: Applying Mechanical Force to Single RNA Molecules Reveals Structure and Function.

In 2010, the Cees Dekker Lab was able to combine the solid-state nanopore platform with optical tweezers to measure the effective force on a dsDNA molecule as it is electrophoretically driven through a SiN nanopore.[56] To combine these systems, DNA molecules were first functionalized with biotin at one end, then attached to a streptavidin-coated polystyrene bead, which is optically trapped by a tightly focused laser beam. When the trapped bead is brought close to a solid-state nanopore, the DNA is pulled into the pore by the emanating electric field. The transmembrane potential can then be varied while the optical tweezers platform is used to report the force on the DNA molecule. Collection of force versus voltage data allows the authors to extract the effective charge of the molecule while confined inside the nanopore. Subsequently, this method has been used to investigate the effects of DNA/protein interaction,[57] localize individual proteins along a DNA strand,[58] and investigate the electroosmotic force and hydrodynamic slip on the nanopore translocation process.[59]

Nynke Dekker Lab has used this system for the study of individual RNA molecules.[51] Using this dual optical tweezers/nanopore system they compared the forces experienced by dsRNA and dsDNA molecules as they translocate a SiN nanopore as a function of the pore diameter (Fig. 5D). Their results indicate that the effective charge of the confined dsRNA is on the order of, but slightly less than the effective charge of dsDNA during the translocation process. Importantly, the authors note that their system is able to make accurate force measurements using nanopores as small as ~3 nm in diameter. This system is of potential interest as it may be adapted to monitor the folding of individual RNA molecules and report the effects of extreme confinement on RNA folding. In addition, this system could be used to detect the presence of individual RNA-binding proteins, monitor their position along a strand and observe conformation changes they induce on RNA molecules.

4.3 Discriminating Individual tRNA Molecules Through Sequential Unfolding

tRNA molecules are good examples of the complex structures that RNA molecules can form. The tRNAs that encode for each amino acid differ slightly from one another, and isodecoder as well as isoacceptor tRNAs add an additional layer of complexity. Their nucleotide sequences and the post-transcriptional modifications that they undergo lead to slight differences in their folded structures. Gerland et al. proposed that by threading structured polynucleotides through an αHL nanopore, it may be possible to sequentially break the intramolecular bonds that they form and "iron out" their secondary structures.[46] They stated that with adequate control over the speed of this process, it should be possible to distinguish the identity of the translocating molecules.

Smith et al. have presented a system that is able to discriminate between two *Escherichia coli* tRNAs[51] by drawing upon the ideas put forth by Gerland et al. The system developed by Smith et al. aims to sequentially unfold and translocate individual tRNA molecules at speeds significantly slower than those observed by simple duplex unzipping experiments. To achieve this, the authors ligate the ends of a tRNA to a synthetic DNA adapter molecule. The 3′ end of the DNA adapter possesses a terminal TEG-linked (triethylene glycol) cholesterol, while the 5′ end possesses an ssDNA leader. The cholesterol on the 3′ end serves to concentrate the molecules at the lipid bilayer, and hence increasing the capture rate of tRNA molecules. This leaves the 5′ end free to be threaded into the nanopore, assisted by the free ssDNA leader. To slow the rate of translocation, the sample is incubated with φ29 DNA polymerase (DNAP), which has been previously used to processively ratchet DNA through a nanopore and read the sequence as each base moves through the pore.[12,13] While this method cannot be used to ratchet the DNA/tRNA hybrid, as the DNAP cannot process the RNA, the authors are able to use an inactive φ29 DNAP as a "molecular brake" to slow the transport. The polymerase preferentially binds near the free 5′ ssDNA end of the DNA/tRNA hybrid, but does not process along the strand, as there are no free DNTPs in the solution. A schematic of what the authors hypothesize the transport process to look like is shown in Fig. 5E. When a voltage bias is applied, any φ29 DNAP adapted DNA/tRNA complexes located near the pore will be electrophoretically captured by the nanopore from their free 5′ ends. From here, the authors partition the translocation process into three steps: (I) translocation of the 5′ end of the adapter beginning with the ssDNA leader, (II) translocation of the tRNA portion, and (III) translocation of the 3′ end of the adapter. The sequence of the DNA adapter contains two abasic residues to assist in identification of these three steps (Fig. 5F). When an abasic residue passes through the pore it results in a significantly larger ionic current, serving as a marker for the strand position, and allowing the authors to identify the part of the signal corresponding to tRNA translocation (II). Characteristic current traces for $tRNA^{fMet}$ and $tRNA^{Lys}$ are shown in Fig. 5F.

Approximately 80 events were collected for each of the tRNAs and a custom support vector machine (SVM) algorithm was employed to use the ionic current parameters contained in each of the three regions in order to distinguish between the two tRNAs. First, each translocation event is analyzed and segregated into the three steps previously mentioned. The mean current and dwell time of each step is then extracted for each translocation event. The SVM plots the extracted mean current and dwell-time values and attempts to draw a line that best separates the points that correspond to the two different tRNAs. In this way, an unknown tRNA can be classified by measuring its dwell time and mean current values and determining on which side of the line this data point resides. The authors find that by analyzing the two adapter regions (I and III), the identity of a translocating tRNA molecule can be classified with accuracies of 60.0 ± 6.9 and 59.4 ± 7.0%, for the two regions respectively. By analyzing only the tRNA region (II) they are able to distinguish the identity of the translocating molecule with an accuracy of 87.2 ± 5.3%.

The inaccuracy of the SVM when applied to the adapter region and the impressive accuracy when applied to the tRNA region, serve as evidence that this method is able to distinguish between two different tRNAs based upon their structural properties. This method employs a sequential bond breaking and unfolding approach to distinguish between molecules, but cannot be used to outright sequence tRNA molecules. In its current incarnation, this technique may have problems scaling up to discriminate more than two tRNAs at a time, but the authors propose that by using an active enzymatic motor that is able to process RNA in single nucleotide steps, they may be able to adapt this method for direct sequencing of tRNA molecules. A nanopore-based system for direct tRNA sequencing would be an important tool for the study of tRNA biology, the effects of posttranslational modifications, and the links between tRNAs and human diseases, adding exciting potential applications for nanopores in RNA studies.

5. CONCLUSIONS

In conclusion, over the past 20 years there have been many studies that evaluate the applicability of nanopores for RNA detection and analysis. Due to their differing chemical structure from their parent DNA, RNA molecules adopt more complex tertiary structures that often resemble protein folding in terms of both structure and function. Both protein-based and solid-state nanopores have been used to study various aspects of RNA molecules, such as their globular (tertiary) structure, sequence, structural dynamics, and their interactions with small molecules and proteins. The further development of nanopore based measurement systems could provide many useful instruments for POC applications and RNA structural studies. Based on the success of single-molecule DNA sequencing using nanopores in recent years, it is logical that a major milestone in nanopore science will be achieved when long-read RNA sequencing is demonstrated. While RNA molecules are typically sequenced using bulk methods by conversion to complementary DNA (cDNA) molecules followed by DNA sequencing, bias errors in library preparation for RNA sequencing and the method's lack of sensitivity toward important RNA modifications are both grand challenges that can be addressed by direct single-molecule RNA sequencing. In the coming years, it will be interesting to see what nanopore-based methods can do in order to facilitate direct RNA sequencing.

REFERENCES

1. Wanunu M. Nanopores: a journey towards DNA sequencing. *Phys Life Rev.* 2012;9: 125–158.
2. Venta K, Shemer G, Puster M, et al. Differentiation of short, single-stranded DNA homopolymers in solid-state nanopores. *ACS Nano.* 2013;7(5):4629–4636.
3. Carson S, Wanunu M. Challenges in DNA motion control and sequence readout using nanopore devices. *Nanotechnology.* 2015;26(7):74004.
4. Branton D, Deamer DW, Marziali A, et al. The potential and challenges of nanopore sequencing. *Nat Biotechnol.* 2008;26(10):1146–1153.
5. Kasianowicz JJ, Brandin E, Branton D, Deamer DW. Characterization of individual polynucleotide molecules using a membrane channel. *Proc Natl Acad Sci USA.* 1996;93 (24):13770–13773.
6. Manrao EA, Derrington IM, Laszlo AH, et al. Reading DNA at single-nucleotide resolution with a mutant MspA nanopore and phi29 DNA polymerase. *Nat Biotechnol.* 2012;30(4):349–353.
7. Cherf GM, Lieberman KR, Rashid H, Lam CE, Karplus K, Akeson M. Automated forward and reverse ratcheting of DNA in a nanopore at 5-Å precision. *Nat Biotechnol.* 2012;30(4):344–348.
8. Dekker C. Solid-state nanopores. *Nat Nanotechnol.* 2007;2(4):209–215.
9. Akeson M, Branton D, Kasianowicz JJ, Brandin E, Deamer DW. Microsecond time-scale discrimination among polycytidylic acid, polyadenylic acid, and polyuridylic acid as homopolymers or as segments within single RNA molecules. *Biophys J.* 1999;77(6): 3227–3233.
10. Butler TZ, Gundlach JH, Troll MA. Determination of RNA orientation during translocation through a biological nanopore. *Biophys J.* 2006;90(1):190–199.
11. Purnell RF, Schmidt JJ. Discrimination of single base substitutions in a DNA strand immobilized in a biological nanopore. *ACS Nano.* 2009;3(9):2533–2538.
12. Stoddart D, Heron AJ, Mikhailova E, Maglia G, Bayley H. Single-nucleotide discrimination in immobilized DNA oligonucleotides with a biological nanopore. *Proc Natl Acad Sci USA.* 2009;106(19):7702–7707.
13. Manrao EA, Derrington IM, Pavlenok M, Niederweis M, Gundlach JH. Nucleotide discrimination with DNA immobilized in the MspA nanopore. *PLoS One.* 2011;6(10): e25723.
14. Ayub M, Bayley H. Individual RNA base recognition in immobilized oligonucleotides using a protein nanopore. *Nano Lett.* 2012;12(11):5637–5643.
15. Ayub M, Hardwick SW, Luisi BF, Bayley H. Nanopore-based identification of individual nucleotides for direct RNA sequencing. *Nano Lett.* 2013;13(12):6144–6150.
16. Clarke J, Wu H-C, Jayasinghe L, Patel A, Reid S, Bayley H. Continuous base identification for single-molecule nanopore DNA sequencing. *Nat Nanotechnol.* 2009;4(4):265–270.
17. Reiner JE, Balijepalli A, Robertson JWF, Drown BS, Burden DL, Kasianowicz JJ. The effects of diffusion on an exonuclease/nanopore-based DNA sequencing engine. *J Chem Phys.* 2012;137:214903.
18. Brady KT, Reiner JE. Improving the prospects of cleavage-based nanopore sequencing engines. *J Chem Phys.* 2015;143:074904.
19. Hunt EA, Broyles D, Head T, Deo SK. MicroRNA detection: current technology and research strategies. *Annu Rev Anal Chem.* 2015;8(1):217–237.

20. Wanunu M, Dadosh T, Ray V, Jin J, McReynolds L, Drndić M. Rapid electronic detection of probe-specific microRNAs using thin nanopore sensors. *Nat Nanotechnol.* 2010;5(11):807–814.
21. Skinner GM, Van Den Hout M, Broekmans O, Dekker C, Dekker NH. Distinguishing single-and double-stranded nucleic acid molecules using solid-state nanopores. *Nano Lett.* 2009;9(8):2953–2960.
22. Wang Y, Zheng D, Tan Q, Wang MX, Gu L-Q. Nanopore-based detection of circulating microRNAs in lung cancer patients. *Nat Nanotechnol.* 2011;6(10):668–674.
23. Zhang X, Wang Y, Fricke BL, Gu L-Q. Programming nanopore ion flow for encoded multiplex microRNA detection. *ACS Nano.* 2014;8(4):3444–3450.
24. Kumar S, Tao C, Chien M, et al. PEG-labeled nucleotides and nanopore detection for single molecule DNA sequencing by synthesis. *Sci Rep.* 2012;2:684.
25. Rodrigues CG, Machado DC, Chevtchenko SF, Krasilnikov OV. Mechanism of KCl enhancement in detection of nonionic polymers by nanopore sensors. *Biophys J.* 2008;95 (11):5186–5192.
26. Huang S, Romero-Ruiz M, Castell OK, Bayley H, Wallace MI. High-throughput optical sensing of nucleic acids in a nanopore array. *Nat Nanotechnol.* 2015;10:986–991. 10.1038/nnano.2015.189.
27. Ramnani P, Gao Y, Ozsoz M, Mulchandani A. Electronic detection of microRNA at attomolar level with high specificity. *Anal Chem.* 2013;85(17):8061–8064.
28. Bayley H, Cronin B, Heron A, et al. Droplet interface bilayers. *Mol Biosyst.* 2008;4(12): 1191–1208.
29. Fischer A, Holden MA, Pentelute BL, Collier RJ. Ultrasensitive detection of protein translocated through toxin pores in droplet-interface bilayers. *Proc Natl Acad Sci USA.* 2011;108(40):16577–16581.
30. Maglia G, Restrepo MR, Mikhailova E, Bayley H. Enhanced translocation of single DNA molecules through alpha-hemolysin nanopores by manipulation of internal charge. *Proc Natl Acad Sci USA.* 2008;105(50):19720–19725.
31. Wanunu M, Bhattacharya S, Xie Y, Tor Y, Aksimentiev A, Drndic M. Nanopore analysis of individual RNA/antibiotic complexes. *ACS Nano.* 2011;5(12):9345–9353.
32. Recht MI, Fourmy D, Blanchard SC, Dahlquist KD, Puglisi JD. RNA sequence determinants for aminoglycoside binding to an A-site rRNA model oligonucleotide. *J Mol Biol.* 1996;262(4):421–436.
33. Fourmy D, Recht MI, Blanchard SC, Puglisi JD. Structure of the A site of *Escherichia coli* 16S ribosomal RNA complexed with an aminoglycoside antibiotic. *Science.* 1996;274 (5291):1367–1371.
34. Pilch DS, Kaul M, Barbieri CM. Ribosomal RNA recognition by aminoglycoside antibiotics. *Anion Sens.* 2005;70(1):179–204. 10.1007/b100447.
35. Niedzwiecki DJ, Iyer R, Borer PN, Movileanu L. Sampling a biomarker of the human immunodeficiency virus across a synthetic nanopore. *ACS Nano.* 2013;7(4):3341–3350.
36. Shasha C, Henley RY, Stoloff DH, Rynearson KD, Hermann T, Wanunu M. Nanopore-based conformational analysis of a viral RNA drug target. *ACS Nano.* 2014;8(6):6425–6430.
37. Xie Y, Dix AV, Tor Y. FRET enabled real time detection of RNA-small molecule binding. *J Am Chem Soc.* 2009;131(48):17605–17614.
38. Morellet N, Jullian N, De Rocquigny H, Maigret B, Darlix JL, Roques BP. Determination of the structure of the nucleocapsid protein NCp7 from the human immunodeficiency virus type 1 by 1H NMR. *EMBO J.* 1992;11(8):3059–3065.
39. Paoletti AC, Shubsda MF, Hudson BS, Borer PN. Affinities of the nucleocapsid protein for variants of SL3 RNA in HIV-1. *Biochemistry.* 2002;41(51):15423–15428.
40. Spahn CM, Kieft JS, Grassucci RA, et al. Hepatitis C virus IRES RNA-induced changes in the conformation of the 40s ribosomal subunit. *Science.* 2001;291(5510): 1959–1962.

41. Seth PP, Miyaji A, Jefferson EA, et al. SAR by MS: discovery of a new class of RNA-binding small molecules for the hepatitis C virus: internal ribosome entry site IIA subdomain. *J Med Chem.* 2005;48(23):7099–7102.
42. Parsons J, Castaldi MP, Dutta S, Dibrov SM, Wyles DL, Hermann T. Conformational inhibition of the hepatitis C virus internal ribosome entry site RNA. *Nat Chem Biol.* 2009;5(11):823–825.
43. Zhou S, Rynearson KD, Ding K, Brunn ND, Hermann T. Screening for inhibitors of the hepatitis C virus internal ribosome entry site RNA. *Bioorg Med Chem.* 2013;21(20): 6139–6144.
44. Rynearson KD, Charrette B, Gabriel C, et al. 2-Aminobenzoxazole ligands of the hepatitis C virus internal ribosome entry site. *Bioorg Med Chem Lett.* 2014;24(15): 3521–3525.
45. Mathé J, Visram H, Viasnoff V, Rabin Y, Meller A. Nanopore unzipping of individual DNA hairpin molecules. *Biophys J.* 2004;87(5):3205–3212.
46. Sauer-Budge AF, Nyamwanda JA, Lubensky DK, Branton D. Unzipping kinetics of double-stranded DNA in a nanopore. *Phys Rev Lett.* 2003;90(23):238101.
47. Gerland U, Bundschuh R, Hwa T. Translocation of structured polynucleotides through nanopores. *Phys Biol.* 2004;1(1–2):19–26.
48. McCauley M, Forties R, Gerland U, Bundschuh R. Anomalous scaling in nanopore translocation of structured heteropolymers. *Phys Biol.* 2009;6(3):036006.
49. Lin J, Kolomeisky A, Meller A. Helix-coil kinetics of individual polyadenylic acid molecules in a protein channel. *Phys Rev Lett.* 2010;104(15):158101.
50. Dewey TG, Turner DH. Laser temperature-jump study of stacking in adenylic acid polymers. *Biochemistry.* 1979;18(26):5757–5762.
51. Van Den Hout M, Vilfan ID, Hage S, Dekker NH. Direct force measurements on double-stranded RNA in solid-state nanopores. *Nano Lett.* 2010;10(2):701–707.
52. Smith AM, Abu-Shumays R, Akeson M, Bernick DL. Capture, unfolding, and detection of individual tRNA molecules using a nanopore device. *Front Bioeng Biotechnol.* 2015;3:91.
53. Moffitt JR, Chemla YR, Smith SB, Bustamante C. Recent advances in optical tweezers. *Annu Rev Biochem.* 2008;77:205–228.
54. Wen J-D, Manosas M, Li PTX, et al. Force unfolding kinetics of RNA using optical tweezers. I. Effects of experimental variables on measured results. *Biophys J.* 2007;92(9): 2996–3009.
55. Frieda KL, Block SM. Direct observation of cotranscriptional folding in an adenine riboswitch. *Science.* 2012;338(6105):397–400.
56. Keyser UF, Koeleman BN, van Dorp S, et al. Direct force measurements on DNA in a solid-state nanopore. *Nat Phys.* 2006;2(7):473–477.
57. Hall AR, Van Dorp S, Lemay SG, Dekker C. Electrophoretic force on a protein-coated DNA molecule in a solid-state nanopore. *Nano Lett.* 2009;9(12):4441–4445.
58. Sischka A, Spiering A, Khaksar M, et al. Dynamic translocation of ligand-complexed DNA through solid-state nanopores with optical tweezers. *J Phys Condens Matter.* 2010;22 (45):454121.
59. Galla L, Meyer AJ, Spiering A, et al. Hydrodynamic slip on DNA observed by optical tweezers-controlled translocation experiments with solid-state and lipid-coated nanopores. *Nano Lett.* 2014;14(7):4176–4182.

CHAPTER FOUR

Photocontrolled Intracellular RNA Delivery Using Nanoparticles or Carrier–Photosensitizer Conjugates

Kazunori Watanabe, Takashi Ohtsuki[1]
Department of Medical Bioengineering, Graduate School of Natural Science and Technology, Okayama University, Okayama, Japan
[1] Corresponding author: e-mail address: ohtsuk@okayama-u.ac.jp

Contents

Abstract

Small interfering RNA (siRNA) and short hairpin RNA (shRNA) may potentially treat a wide variety of diseases through RNA interference-mediated silencing of specific genes. MicroRNAs (miRNAs) are endogenous regulatory RNA molecules that also modulate gene expression, and thus potential therapeutic approaches using miRNAs have attracted attention. For clinical application of these small RNAs, efficient and safe RNA delivery to target tissues and cells is necessary. Current challenges to RNA delivery are the penetration of negatively charged RNAs through the cell membrane and specific delivery of RNA into target cells. Photocontrolled intracellular RNA delivery is a promising strategy with high target specificity. This strategy includes photodependent endosomal escape of RNA or photodependent release of RNAs from carrier particles. In this chapter, photocontrolled intracellular RNA delivery methods employing gold or silver nanoparticles, upconversion nanoparticles, proteins, or polymers are discussed.

Progress in Molecular Biology and Translational Science, Volume 139
ISSN 1877-1173
http://dx.doi.org/10.1016/bs.pmbts.2015.10.013

1. INTRODUCTION

Small RNAs, especially, those inducing specific gene silencing via RNA interference (RNAi), are being widely explored as a therapeutic strategy for intractable diseases, including cancer, neurodegenerative diseases, viral infections, and as a tool for functional analysis.[1]

Small interfering RNAs (siRNAs) and short hairpin RNAs (shRNAs), which are completely complementary to the target mRNA, are functional mediators of RNAi.[2–4] An siRNA or shRNA is incorporated into the RNA-induced silencing complex (RISC), which then binds and degrades a target mRNA. MicroRNAs (miRNAs), in contrast to siRNAs and shRNAs, are able to recognize the target mRNA using 6–8 nucleotides at 5′ end of the miRNA.[5,6] The mechanism for miRNA-mediated gene silencing resembles that of RNAi-mediated silencing in that an Argonaute protein-associated RNA (siRNA antisense strand or miRNA) binds to a specific mRNA and interferes with its translation.[7]

For studies using small RNAs, development of an effective and nontoxic RNA delivery system is necessary.[8–10] Although viral vector technologies can stably introduce RNA genes into the host genome, they are associated with many potential side effects including immunotoxicity and mutagenesis.[11–13] In recent years nonviral delivery systems using various nonviral agents, such as lipid-based agents,[14,15] cationic polymers,[16] and cell-penetrating peptides [CPPs, also called peptide transduction domains (PTDs)],[17–20] have been studied. Lipid-based agents and cationic polymers are used as standard tools for RNA transfection in vitro because of their relatively low production costs and low immunogenicity[21,22]; although, these agents cannot specifically deliver RNAs to target cells. Local delivery of siRNA or shRNA into target cells is required to reduce side effects associated with RNAi therapeutics. Thus the development of methods for spatially and temporally controlled RNA delivery is desired.

Recently, many groups have attempted to use light for the spatial and temporal control of intracellular delivery of biologically active macromolecules such as drugs, proteins, and nucleic acids.[23–26] Compared to chemical inducers, light is a more favorable inducer for fine control of intracellular molecular delivery. In addition, photocontrolled molecular delivery methods can be applied in photodynamic therapy (PDT), a strategy to eradicate local tumors photodependently.[27,28] Photocontrol of intracellular RNA delivery is typically performed by the photochemical internalization (PCI) strategy using photosensitizers (PSs) or metal nanoparticles that can absorb near-infrared (NIR) light at 800–1000 nm.[29] There are several reports about RNA delivery by the PCI strategy of using an RNA carrier and a PS separately;[30–35] however, here, we focus on RNA delivery by using photo-responsive RNA carriers, which have photodependent cytosolic RNA delivery ability or photodependent RNA release ability. The advantage using nanoparticles as carriers is that these particles can be delivered specifically to tumor sites by the enhanced permeability and retention (EPR) effect. In this chapter, we discuss photocontrolled RNA delivery methods employing gold nanoparticles (AuNPs) or silver nanoparticles (SNPs), upconversion nanoparticles (UCNPs), proteins, or polymers.

2. GOLD NANOPARTICLES

AuNPs have emerged as versatile, selective, and multifunctional anticancer therapeutics because of their unique properties and represent a powerful alternative to conventional RNA delivery systems.[36] The first demonstration of gene silencing with DNA-AuNPs by an antisense mechanism, not by the RNAi mechanism, was reported by Rosi and coworkers.[37] Furthermore, gene silencing using siRNA-AuNPs was reported by Oishi et al. and Giljohann et al.[38,39] These reports demonstrated nucleic acid delivery using AuNPs, but did not demonstrate photodependent RNA delivery.

Nanostructures of noble metals, such as gold and silver, exhibit unique optical properties due to strong surface plasmon resonance absorption at NIR wavelength, which can present photothermal effects to trigger a variety of biological activities. For example, Hirsch et al. demonstrated that gold nanoshell (NS) treatment followed by NIR irradiation induced cell death.[40] A gold NS is an AuNP with a hollow interior. Wijaya et al. demonstrated photodependent melting of gold nanorods (Au NRs), which released thiol-conjugated DNA oligonucleotides.[41]

Braun et al. demonstrated temporally and spatially controlled RNAi using AuNP-conjugated siRNA (Fig. 1).[42] They employed unique combination of features of AuNPs: small size (a diameter of ~40 and ~3 nm thick shells), hollow interior, and tunable absorption band. The surfaces of AuNPs were modified with polyethylene glycol (PEG)-thiolated siRNA against enhanced green fluorescence protein (EGFP) and coated with a Tat peptide-lipid for cellular uptake of the nanoparticle at a picomolar concentration. EGFP

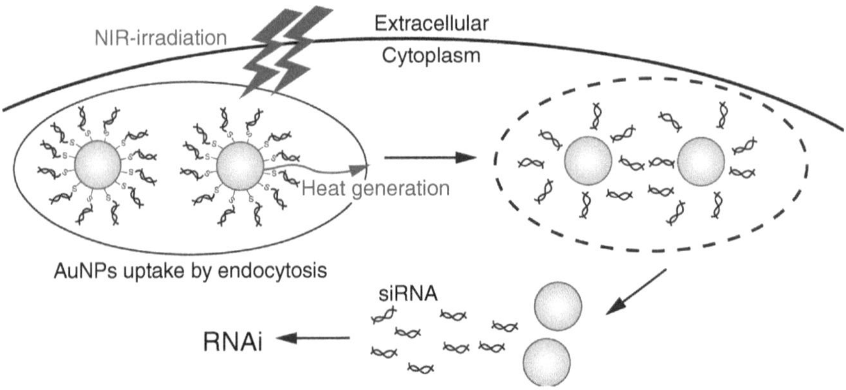

Figure 1 Schematic illustration of photoinduced RNAi using AuNPs. The endosomally entrapped siRNA-conjugated AuNPs can escape from the endosomes by NIR irradiation because of NIR-induced heat generation from the AuNPs. The released siRNAs induce RNAi in the cytoplasm.

expression in C166-GFP cells treated with modified AuNPs was ~80% reduced by NIR laser irradiation [800 nm, 105 J/cm^2 (3.5 W/cm^2, 30 s)]. In addition, the gene silencing of EGFP was confined to the NIR laser/AuNPs treatment area.

More recently, several researchers also reported NIR-responsive gene silencing by using siRNA-carrying AuNPs. Au NRs with siRNA (or oligonucleotide)-capped mesoporous silica ($mSiO_2$) shells were employed by Chang et al. (Fig. 2A).[43] A continuous-wave diode laser (808 nm), which is inexpensive and easy to handle, was used to irradiate the Au NR-$mSiO_2$ (AuMS) hybrid containers. The thiolated siRNA (or oligonucleotide) was conjugated with AuMS by using 4-maleimidobutyric acid *N*-succinimidyl ester as a linker. The double-stranded oligonucleotide-modified AuMS showed NRs with ~41 nm length and ~11 nm width covered by a ~16 nm thick SiO_2 shell. After GFP-expressing HeLa cells were incubated with the GFP-targeting siRNA-AuMS and were irradiated with the laser [808 nm, 900 J/cm^2 (1.5 W/cm^2, 10 min)], the expression of GFP was decreased to ~33% at 72 h. Chang et al. also prepared doxorubicin-loaded AuMS. Photodependent decrease in viability of A549 lung cancer cells was observed in the presence of the doxorubicin-loaded AuMS.

Huschka et al. employed a gold NS conjugated with a PLL peptide, which associated electrostatically with the negatively charged DNA antisense oligonucleotide or siRNA (Fig. 2B).[44] PLL can protect synthetic oligonucleotide-based molecules from degradation by nucleases and can increase

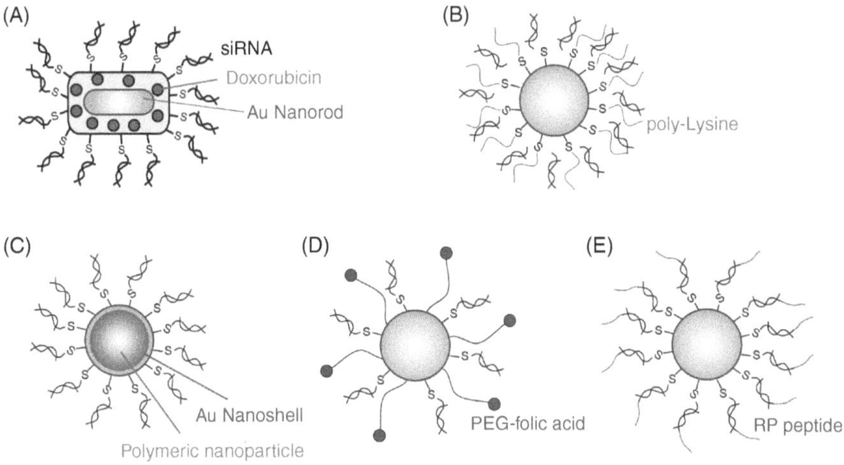

Figure 2 Schematic illustrations of AuNPs. (A) Au NR with siRNA-capped mesoporous silica shell loaded with doxorubicin.[42] (B) Gold NS conjugated with poly(L-lysine) (PLL) peptides and electrostatically associated negative charged siRNAs.[44] (C) Polymer gold nanoshell (PGN) coated with pyrogallol 2-aminoethane (PAE) carrying siRNAs.[45] (D) Hollow gold nanosphere (HAuNS) carrying PEG–folic acid conjugates and siRNAs.[46] (E) Hollow gold nanoshell (HGN) coated with cancer cell-targeting peptide, RPARPAR (RP).[47]

their cellular uptake. They demonstrated GFP knockdown by using the NS-PLL in H1299 lung cancer cells. After incubation of GFP-expressing H1299 cells for 12 h with NS-PLL-ssDNA and -siRNA, the GFP expression was downregulated to 72 ± 5 and 80 ± 5%, respectively. The cells were irradiated with the laser [800 nm, 300 J/cm^2 (2.5 W/cm^2, 2 min)] after the 12-h incubation and at 6 h after irradiation, GFP expression was downregulated to ~47% by antisense ssDNA and ~49% by siRNA. Incubation with NS-PLL accompanied by laser treatment resulted in 88 ± 7% cell survival, indicating that cytotoxicity induced by NS-PLL and laser irradiation was minimal.

Jeong et al. prepared PGNs by a novel plant-inspired flavonoid surface modification method (Fig. 2C).[45] PGNs were prepared using polymeric nanoparticles made of polymethylmethacrylate. The nanoparticle was coated with PAE, which enables facile metal shell formation, such that a thin gold layer was plated on the PAE-coated polymeric nanoparticles. Finally, the PGNs were conjugated with thiolated siRNAs by thiol-Au interactions. GFP-expressing HeLa cells that had been treated with PGN–siRNA conjugates (100 nM siRNA) were irradiated with the laser [808 nm, 450 J/cm^2 (1.5 W/cm^2, 5 min)]; consequently, 51.7% silencing of GFP expression was achieved. Cytotoxicity was not observed after 10 min exposure to light intensities of both 1.5 and 2.4 W/cm^2. However, a slight increase in cytotoxicity was observed after 20 min exposure to light intensity of 2.4 W/cm^2.

Target cell-specific gene silencing in vitro and in vivo via NIR-induced siRNA release from AuNPs was reported by Lu et al.[46] They employed HAuNSs carrying folic acids through thioctic acid–terminated PEG linkers and thiolated siRNA (F-PEG-HAuNS-siRNAs) (Fig. 2D). The F-PEG-HAuNS-siRNAs specifically targeted folate receptor-positive cells through the folic acid and were internalized by cells via an endocytotic pathway. Following NIR irradiation [800 nm, 3 J/cm^2 (50 mW/cm^2, 60 s)], F-PEG-HAuNS-siRNAs targeting NF-κB p65 were released into the cytoplasm of HeLa cells that express high levels of folate receptors; consequently, the expression of NF-κB p65 was decreased by ~95% at 72 h after the irradiation. NIR-induced tumor-specific RNAi was also demonstrated in nude mice bearing HeLa cervical cancer xenografts. NF-κB p65 expression was reduced to 23% after NIR irradiation. In addition, compared to treatment with irinotecan alone, combined treatment with irinotecan and F-PEG-HAuNS-siRNA with NIR irradiation enhanced the apoptotic response in folate receptor-positive cells. Consequently, tumor growth was delayed by the combined treatment.

The target-cell-specific gene silencing via NIR-induced siRNA release from HGNs coated with cancer cell–targeting peptide, RPARPAR (RP), was reported by Huang et al (Fig. 2E).[47] RP binds specifically to cancer cells that overexpress the neuropilin-1 receptor, such as PPC-1 cells. Polo-like kinase (PLK1) was targeted to induce a proapoptotic pathway. Several phase I/II clinical trials for the treatment of solid tumors have been based on RNAi-based PLK1 downregulation. Femtosecond-pulsed NIR laser irradiation [800 nm, 24 J/cm^2 (2.4 W/cm^2, 10 s)] of PPC-1 cells exposed to HGNs conjugated with RP and siRNA (HGN-SD-RP) led to a 70% decrease in cell viability, which is similar to that observed after siRNA transfection using Lipofectamine RNAiMAX. Cells exposed to the same laser conditions in the absence of HGN-SD-RP showed no loss of viability. HGN-SD-RP followed by the NIR treatment decreased the PLK1 expression in PPC-1 cells to ~30–40% at 72 h after irradiation. To confirm cell-specific targeting, the authors attempted to introduce HGN-SD-RP into RWPE-1 cells lacking the neuropilin-1 receptor on the cell surface; no downregulation of PLL-1 or loss of cell viability was observed after the laser treatment.

3. SILVER NANOPARTICLES

Instead of using AuNPs, some groups have recently developed photoinduced RNAi using SNPs.[48,49] SNPs have unique optical properties. The light scattering cross-section of an SNP is ~10 times greater than that of a similarly sized AuNP. In addition, compared to other metallic nanomaterials, SNPs offer higher extinction coefficients and higher blue-shifted plasmon resonant peaks,[50,51] making them an appropriate alternative for photoinduced drug delivery.

Brown et al. demonstrated photoinduced gene silencing by using SNP-carrying photolabile DNA oligonucleotides,[48] although RNA delivery was not demonstrated in this study. SNPs and oligonucleotides were connected by a nitrophenylethyl photocleavable (PC) linker, which initiates disengagement of the oligonucleotide when exposed to UV light. Transmission electron microscopy analysis revealed the diameter of SNP-oligonucleotides as 74 ± 21 nm. Oligonucleotides connected to SNPs were effectively protected from DNase I digestion. In vitro analysis showed that TYE563-labeled oligonucleotides were released from SNPs by low (302 nm, 5.2 mW/cm^2) and high (365 nm, 206 mW/cm^2) wavelength irradiation. HeLa cells treated with SNP-PC-oligonucleotides targeting intercellular adhesion molecule-1 (ICAM-1, CD54) were irradiated with the UV laser (365 nm), and CD54 gene silencing was achieved. Cell viability and CD54 expression level were not affected by the UV irradiation.

Qureshi et al. demonstrated spatial and temporal modulation of osteogenesis in human autologous adipose-derived mesenchymal stromal/stem cells (hASCs) by using SNPs conjugated with microRNA-148b mimics (miR-148b) via nitrobenzyl PC linkers (Fig. 3).[49] This miRNA induces an increase in total alkaline phosphatase (ALP) activity, which is an early

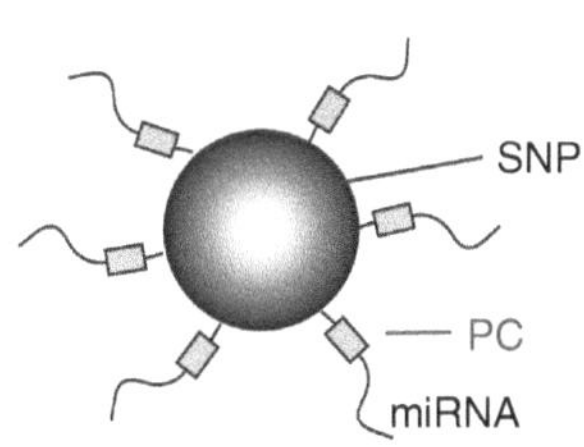

Figure 3 SNP conjugated with miRNAs via PC linkers.

biomarker of osteogenesis in bone marrow–derived mesenchymal stem cells.[52] The size of the photocleavable miR-148b/SNP conjugate (PC-miR-148b/SNP) was 75.37 ± 0.83 nm. Increase in ALP activity was observed at day 7, after hASCs were treated with PC-miR-148b/SNP and UV irradiation (365 nm, 52.4 J/cm^2). Upregulation of ALP induced by PC-miR-148b/SNP was similar to that in hASCs treated with osteogenic media. Expressions of mRNA middle and late stage osteogenic markers, RunX2 and osteocalcin, respectively, were significantly upregulated at days 7 and 28, after photoactivation of PC-miR-148b/SNP in cells. In addition, an indication of bone formation, mineralization, was induced by PC-miR-148b/SNP and UV irradiation. Cytotoxicity induced by PC-miR-148b/SNP and UV treatment was minimal; the treatment resulted in 90.08 ± 2.12% cell survival.

4. UPCONVERSION NANOPARTICLES

UCNPs have been studied for applications in PDT and imaging of biomolecules.[53–55] UCNPs can absorb long wavelength NIR light and release narrow and sharp emissions ranging from UV to visible light. NIR light penetrates much more deeply into tissue compared to UV to visible light. However, most of caged compounds can be activated by UV to visible light, not by NIR light. Thus UCNPs that emit UV to visible light by absorbing NIR light are advantageous in activating caged compounds, such as caged siRNAs, in deep tissue.

Zhang et al. showed that treatments with NIR and antibody (anti-MUC1/episialin)-carrying UCNPs molecules induced death of MCF-7/AZ human mammary carcinoma cells that highly expressed MUC1/episialin.[53] Jiang and Zhang demonstrated cytosolic siRNA delivery using UCNPs.[56] Although their study did not include acceleration of cytosolic siRNA release by NIR, intracellular fate of the siRNA was shown using fluorescence imaging.

Recently, several groups have developed RNA delivery into the cytoplasm using UCNP–siRNA conjugates and NIR irradiation.[57–61] In 2012 Jayakumar et al. utilized UCNPs coated with mesoporous silica, in which caged siRNAs were loaded, for NIR-dependent RNAi (Fig. 4A). The UCNPs were made of sodium yttrium fluoride ($NaYF_4$), ytterbium (Yb^{3b}), and thulium (Tm^{3b}). When UCNPs are irradiated by NIR light, they emit UV-visible light. The caged siRNA, siRNA protected with

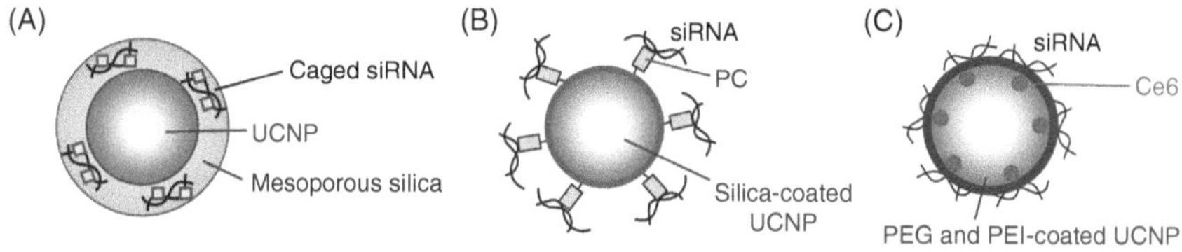

Figure 4 UCNPs. (A) UCNPs coated with mesoporous silica, in which caged-siRNAs are loaded.[57,58] (B) Silica-coated UCNP carrying siRNAs via PC linker.[59] (C) PEG- and branched PEI-coated UCNPs loaded with siRNAs and chlorin e6 (Ce6).[61]

4,5-dimethoxy-2-nitroacetophenone (DMNPE), can be uncaged by UV light. In vitro analysis showed that the caged siRNA molecules on the UCNPs were activated under excitation by NIR laser at 980 nm. There was a significant decrease in GFP fluorescence in B16F0 mouse melanoma cells after caged anti-GFP siRNA delivery and 1344 J/cm^2 of NIR laser exposure when compared to unirradiated controls. No cytotoxicity was observed in B16F0 cells treated with UCNPs/DMNPE-siRNA and NIR laser.

Guo et al. performed a gene therapy model experiment using UCNPs/DMNPE-siRNA. They attempted to induce silencing of the survivin gene and to inhibit growth of murine bladder cancer cells (MB49 cells).[58] Survivin, a strong apoptosis inhibitor, has been used as a target for anticancer therapy and as a tumor marker.[62,63] Monodisperse silica-coated UCNPs based on $NaYF_4$, Yb^{3b}, and Tm^{3b} were attached to DMNPE-siRNA through electrostatic interaction. In MB49 cells treated with 500 μg/mL UCNPs/DMNPE-siRNA followed by NIR irradiation [980 nm, 90 J (100 mW, 15 min)], the expression of survivin decreased to 8.8% at 24 h after irradiation compared to that in cells without irradiation (29.8%). The viability of MB49 cells was considerably decreased (to ~20%) with NIR laser (100 mW, 15 min), although it was also decreased (to ~70%) without NIR laser.

Yang et al. presented a new method that combined UCNPs and PC linkers (Fig. 4B).[59] They synthesized a silica-coated UCNP ($NaYF_4$, Yb^{3b}, and Tm^{3b}) conjugated with photoactive *o*-nitrobenzyl ester-functionalized thiol linker, with a positively charged alkyl amine at the end of the structure. The UCNPs associated with anionic siRNAs through electrostatic interactions. The UCNP–siRNA complex could be photocleaved and the adsorbed siRNAs were easily released from surface of UCNPs by NIR irradiation. After HeLa cells were pretransfected with EGFP plasmid, treated with the UCNPs–siRNA complex, and irradiated by a 980-nm diode laser (2 h, with a 5-min break after each 30-min exposure), the EGFP expression was significantly decreased. Light intensity was not clearly described. Cell viability after UCNP/siRNA and NIR treatment was more than 80%, which is similar to that in cells treated with Lipofectamine 2000.

Jayakumar et al. presented a combination of UCNPs: a photomorpholino and a PS TPPS2a (*meso*-tetraphenylporphine with two sulfonate groups on adjacent phenyl rings).[60] They synthesized mesoporous silica-coated UCNPs (based on $NaYF_4$, Yb^{3+}, Tm^{3+}, and Er^{3+}) associated with TPPS2a and an anti-STAT3 photomorpholino (double-stranded: a sense morpholino with a photolabile moiety and an unmodified antisense). Their concept was that NIR-excited UCNPs simultaneously emit both UV and visible light, causing excitation of TPPS2a by visible light at 413 nm and activation of the photomorpholino by UV light. UCNP/TPPS2a/photomorpholino particles were entrapped in endosomes when B16F0 cells were incubated with the particles, but NIR irradiation [980 nm, 1344 J/cm^2 (2.8 W/cm^2, 8 min)] improved endosomal escape of the particles compared to UCNPs only. This endosomal escape may be due to reactive oxygen species generated from the excited PS TPPS2a. UCNPs/TPPS2a/anti-STAT3 photomorpholinos and NIR irradiation decreased the expression of STAT3 to ~50% in B16F0 cells, although STAT3 also decreased (to ~80%) after treatments with NIR irradiation and UCNPs/anti-STAT3 photomorpholinos lacking TPPS2a. The cytotoxicity with UCNPs/TPPS2a/anti-STAT3 photomorpholinos was similar to that with UCNPs/anti-STAT3 photomorpholinos. In vivo analysis showed that the expression of STAT3 was decreased by ~60% in tumor tissues treated with UCNPs/TPPS2a/anti-STAT3 photomorpholinos and NIR irradiation, compared with in-control tissues treated with PBS and NIR irradiation. Furthermore, the treatment with UCNPs/TPPS2a/anti-STAT3 photomorpholinos and NIR irradiation induced the reduction of tumor volume in mice compared to the treatment with PBS without NIR irradiation.

Wang et al. synthesized UCNPs (based on $NaGdF_4$, Yb, and Er) with multilayered polymer coatings [PEG and branched polyethylenimine (PEI)] for siRNA delivery (Fig. 4C).[61] Both siRNA and PS Ce6 molecules were loaded into the coated UCNPs. The target for the siRNA was PLK1. Silencing of PLK1 triggers apoptosis, which may be applied to cancer therapy. Treatment with UCNPs/Ce6/siRNA and NIR irradiation (980 nm, 0.8 W/cm^2 for 20 min with an interval for every 1 min of light exposure to avoiding heating) induced downregulation of PLK1 mRNA to ~35% in HeLa cells. UCNPs/Ce6/siRNA induced the decrease of cell viability to ~20% in the presence of NIR laser and ~60% in the absence of NIR irradiation. UCNPs/Ce6, which did not include the siRNA, reduced cell viability to ~50% in the presence of NIR irradiation.

5. PROTEIN–PHOTOSENSITIZER CONJUGATES

Instead of using metals, such as gold, silver, or yttrium, photodependent RNA delivery was also achieved using an RNA binding protein (RBP) carrying a PS (Fig. 5). In 2008 our group demonstrated photodependent intracellular RNA delivery using TatU1A protein.[64,65] TatU1A consists of a Tat CPP and an RBP derived from the U1A RNA-binding domain (RBD), which recognizes a short RNA sequence (5′-CAUUGCACUCCG-3′) with high specificity and affinity.[66] The Tat peptide is known to deliver various cargos into animal cells, mainly via endocytosis. Although any RNA bearing the U1A-recognized sequence could be delivered by TatU1A, the RNA and TatU1A were trapped in the endosomes. A PS was attached to the C-terminus of TatU1A for photodependent endosomal escape thorough the PCI mechanism. Treatments with TatU1A-PS/shRNA (anti-GFP) complexes and visible light irradiation (at 540 ± 10 nm, 10 J/cm^2) decreased EGFP expression to ~30% in Chinese hamster ovary (CHO) cells.[67] In addition, spatial control of EGFP gene silencing was demonstrated using a photomask (approx. millimeter scale) and a confocal laser (approx. micrometer scale).[68] No cytotoxicity was observed by this method in CHO cells. We designated this photoinduced RNAi strategy as the CPP-linked RBP-mediated RNA internalization and photoinduced RNAi (CLIP-RNAi) method. Other CPPs and RBPs were also employed and five additional CPP-RBP variants were constructed.[67] Among them, the most efficient carrier proteins were TatU1A-CL1 bearing the degradation signal peptide CL1 and FHVU1A bearing a flock house virus (FHV) coat peptide as a CPP. Photoinduced RNAi was also achieved using TatU1A-CL1-PS and

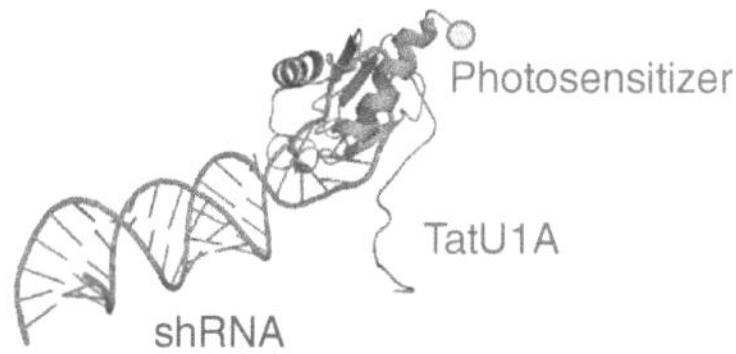

Figure 5 TatU1A–PS conjugate and an shRNA. The accession numbers of the structures used here are 1R9F (siRNA) (a part of the shRNA in this figure) and 1U6B (U1A).

FHVU1A-PS. Attachment of CL1 to TatU1A accelerated RNAi-mediated gene silencing, suggesting that dissociation of the carrier protein and RNA caused by degradation of the carrier in the cytosol is important for RNAi activity. Furthermore, we examined the various PS to expand the excitation wavelength of the CLIP-RNAi method to the NIR region. The complex of TatU1A carrying DY750 as a PS (TatU1A-DY750) and shRNA downregulated EGFP expression in EGFP-CHO cells by NIR irradiation (735–765 nm, 45 J/cm^2).[69] Photoinduced singlet oxygen generation from TatU1A-DY750 was significantly higher than that from other candidate TatU1A-PS molecules that can absorb NIR, suggesting that singlet oxygen is important for disruption of the endosome. These photodependent mechanisms for cytosolic RNA delivery and RNAi suggest new possibilities for cellular and molecular biology and therapy.[29]

6. POLYMER COMPLEXES CONTAINING PHOTOSENSITIZERS OR PHOTODEGRADABLE LINKERS

Recently, the photodependent RNA delivery method using synthetic polymers have been reported.[70–72] Park et al. demonstrated a decrease in epidermal growth factor receptors (EGFR) in HCT116 cells (colon carcinoma cell line) using nanoparticles consisting of positively charged PEI, acetylated chondroitin sulfate (CS) modified with PSs [pheophorbide a (PhA)], and anti-EGFR shRNA.[70] CS was employed to achieve cancer-specific targeting because it has been reported to be internalized by cancer cells via CD44 receptor-mediated endocytosis. The diameter of PEI-PhA-CS-shRNA was 90 nm. Treatment of HCT116 cells that highly express CD44 receptors with PEI-PhA-CS-shRNA and irradiation (633 nm, 0.5 J/cm^2) induced EGFR gene silencing up to ~85%. In addition, the combination of PEI-PhA-CS-shRNA and photoirradiation (30 J/cm^2) inhibited the tumor growth in HCT116 tumor-bearing BALB/c nude mice.

Ma et al. used a star-shaped polymer (PP-PLLD) consisting of a porphyrin (PP) core and arginine-modified 3rd generation PLL dendron arms.[71] Porphyrin was used as a PS. PP-PLLD-Arg could bind to hydrophobic docetaxel (DOC) because of its amphiphilic property and could bind to anti-MMP-9 shRNA plasmid through electrostatic interactions. The diameter of PP-PLLD-Arg/DOC/MMP-9 complexes was 150 nm. PP-PLLD-Arg/MMP-9 complexes were protected from DNase I digestion in vitro and

PP-PLLD-Arg had better blood compatibility and lower cytotoxicity compared to PEI in vivo. After HNE-1 cells were treated with PP-PLLD-Arg/MMP-9 complexes, expressions of MMP-9 mRNA and protein were decreased by ~20% compared to that in cells treated with PBS. After irradiation (435 nm, 1.26 J/cm^2), MMP-9 protein expression was about 50% decreased in the cells treated with PP-PLLD-Arg/MMP-9 complexes. In addition, codelivery of DOC and anti-MMP-9 shRNA plasmid by the polymer complex could induce cell apoptosis effectively.

Murayama et al. reported photoinduced RNAi using siRNA encapsulated in photodegradable nanoparticles,[72] instead of nanoparticles with PSs. They synthesized nanoparticles based on photodegradable PEG-Photo-acryloyl chloride (Ac) and DAB-Ac. PEG-Photo-Ac imparts photodegradability to the nanoparticles. DAB-Ac facilitates internalization of the particles by cells and binds to siRNA (antiluciferase) through electrostatic interaction. The diameter of PEG-Photo-Ac/DAB-Ac/siRNA was 120 nm. In vitro analysis showed that irradiation induced siRNA release from the nanoparticles. The siRNA-containing nanoparticles were internalized by human ovarian cancer cells (SKOV3-Luc), and the siRNAs were released by UV irradiation for 3 min. The activity of luciferase was decreased to 40% in UV irradiated cells compared to unirradiated cells.

7. CONCLUSIONS

Spatially and temporally controlled intracellular RNA delivery is a significant challenge for applying RNAs in clinically useful therapies. One approach to overcome this challenge is the development of technologies for facilitating photodependent intracellular RNA delivery. To accomplish photodependent RNA delivery, many research groups have devised photoresponsive RNA carriers such as metal nanoparticles, polymer-based particles, and photosynthesizer-carrying proteins (Table 1). NIR irradiation-induced cellular RNA internalization was achieved by several nanoparticles and the protein–PS conjugate. NIR-stimulation is favorable for in vivo application because tissue absorption of light in the NIR region is minimal.[73] Several AuNPs and UCNPs and polymer–PS conjugates could induce cancer cell death through RNAi and PDT. Thus these methods are promising alternatives for gene therapy and PDT; further advances in the design of photoresponsive RNA carriers are expected in future.

Table 1 Nanoparticles and Carrier-Photosensitizer Conjugates

Nanoparticle	Size (nm)	Wavelength (nm)	Energy Dose (J/cm²)	Cell Line	Target	RNAi Efficiency (%)	References
Gold nanoparticles							
AuNPs	~40	800	105	C166-GFP	EGFP	~80	[42]
AuMS	~41 (length) ~11 (width)	808	900	HeLa	GFP	~67	[43]
Gold NS conjugated with PLL		800	300	H1299	GFP	51	[44]
PGNs	111	808	450	HeLa	GFP	51.7	[45]
F-PEG-HAuNS-siRNAs		800	3	HeLa	p65	95	[46]
		800		HeLa in mice	p65	23	
HGNs-SD-RP		800	24	PPC-1	PLK1	~70	[47]
Silver nanoparticles							
SNP-PC-oligonucleotide	74	365	25	HeLa	CD54	~70	[48]
PC-miR-148b/SNP	75.4	365	52.4	hASCs		not RNAi	[49]
Upconversion nanoparticles							
UCNPs-DMNPE-siRNA	~80	980	1344	B16F0	GFP	~35	[57]
UCNPs/DMNPE-siRNA		980	(90 J)	MB49	Survivin	93	[58]

UCNP-siRNA	~74	980		HeLa	EGFP	~90	[59]
UCNPs/TPPS2a/anti-STAT3	~30 (core shell)	980	1344	B16F0	STAT3	~50	[60]
UCNPs/Ce6/siRNA	~30 (UCNPs)	980	960	HeLa	PLK1	~60	[61]
Protein–photosensitizer conjugates							
TatU1A-PS/shRNA		530–550	10	CHO	EGFP	~70	[67]
TatU1A-DY750/shRNA		735–765	45	CHO	EGFP	70	[69]
Polymer complex containing photosensitizer or photodegradation linker							
PEI-PhA-CS-shRNA	90	633	0.5	HCT116	EGFR	~80	[70]
PP-PLLD-Arg/MMP-9	150 (PP-PLLD-Arg/DOC/MMP-9)	435	1.26	HNE-1	MMP-9	~50	[71]
PEG-Photo-Ac/DAB-Ac/siRNA	120	UV	—	SKOV-3	Luciferase	60	[72]

REFERENCES

1. Pecot CV, Calin GA, Coleman RL, Lopez-Berestein G, Sood AK. RNA interference in the clinic: challenges and future directions. *Nat Rev Cancer.* 2011;11:59–67.
2. Fire A, Xu S, Montgomery MK, et al. Potent and specific genetic interference by double-stranded RNA in *Caenorhabditis elegans*. *Nature.* 1998;391:806–811.
3. Zamore PD, Tuschl T, Sharp PA, Bartel DP. RNAi: double-stranded RNA directs the ATP-dependent cleavage of mRNA at 21 to 23 nucleotide intervals. *Cell.* 2000;101:25–33.
4. Elbashir SM, Harborth J, Lendeckel W, et al. Duplexes of 21-nucleotide RNAs mediate RNA interference in cultured mammalian cells. *Nature.* 2001;411:494–498.
5. Lewis BP, Shih IH, Jones-Rhoades MW, Bartel DP, Burge CB. Prediction of mammalian microRNA targets. *Cell.* 2003;115:787–798.
6. Bartel DP. MicroRNAs: target recognition and regulatory functions. *Cell.* 2009;136:215–233.
7. Hutvagner G, Simard MJ. Argonaute proteins: key players in RNA silencing. *Nat Rev Mol Cell Biol.* 2008;9:22–32.
8. Gooding M, Browne LP, Quinteiro FM, Selwood DL. siRNA delivery: from lipids to cell-penetrating peptides and their mimics. *Chem Biol Drug Des.* 2012;80:787–809.
9. Shim MS, Kwon YJ. Efficient and targeted delivery of siRNA in vivo. *FEBS J.* 2010;277:4814–4827.
10. Williford JM, Wu J, Ren Y, et al. Recent advances in nanoparticle-mediated siRNA delivery. *Annu Rev Biomed Eng.* 2014;16:347–370.
11. Raper SE, Chirmule N, Lee FS, et al. Fatal systemic inflammatory response syndrome in a ornithine transcarbamylase deficient patient following adenoviral gene transfer. *Mol Genet Metab.* 2003;80:148–158.
12. Check E. Gene therapists urged to learn more immunology. *Nature.* 2005;434:812.
13. Guo J, Evans JC, O'Driscoll CM. Delivering RNAi therapeutics with non-viral technology: a promising strategy for prostate cancer? *Trends Mol Med.* 2013;19:250–261.
14. Foged C. siRNA delivery with lipid-based systems: promises and pitfalls. *Curr Top Med Chem.* 2012;12:97–107.
15. Rehman Z, Zuhorn IS, Hoekstra D. How cationic lipids transfer nucleic acids into cells and across cellular membranes: recent advances. *J Control Release.* 2013;166:46–56.
16. Nimesh S, Gupta N, Chandra R. Cationic polymer based nanocarriers for delivery of therapeutic nucleic acids. *J Biomed Nanotechnol.* 2011;7:504–520.
17. Fonseca SB, Pereira MP, Kelley SO. Recent advances in the use of cell-penetrating peptides for medical and biological applications. *Adv Drug Deliv Rev.* 2009;61:953–964.
18. van den Berg A, Dowdy SF. Protein transduction domain delivery of therapeutic macromolecules. *Curr Opin Biotechnol.* 2011;22:888–893.
19. Nakase I, Tanaka G, Futaki S. Cell-penetrating peptides (CPPs) as a vector for the delivery of siRNAs into cells. *Mol Biosyst.* 2013;9:855–861.
20. Endoh T, Ohtsuki T. Cellular siRNA delivery using cell-penetrating peptides modified for endosomal escape. *Adv Drug Deliv Rev.* 2009;61:704–709.
21. Akhtar S, Benter IF. Nonviral delivery of synthetic siRNAs in vivo. *J Clin Invest.* 2007;117:3623–3632.
22. Park TG, Jeong JH, Kim SW. Current status of polymeric gene delivery systems. *Adv Drug Deliv Rev.* 2006;58:467–486.
23. Mikat V, Heckel A. Light-dependent RNA interference with nucleobase-caged siRNAs. *RNA.* 2007;13:2341–2347.
24. Yu H, Li J, Wu D, Qiu Z, Zhang Y. Chemistry and biological applications of photo-labile organic molecules. *Chem Soc Rev.* 2010;39:464–473.
25. Zhao Y, Biggs TD, Xian M. Hydrogen sulfide (H2S) releasing agents: chemistry and biological applications. *Chem Commun.* 2014;50:11788–11805.

26. Yang D, Ma P, Hou Z, et al. Current advances in lanthanide ion (Ln(3+))-based upconversion nanomaterials for drug delivery. *Chem Soc Rev.* 2015;44:1416–1448.
27. Benov L. Photodynamic therapy: current status and future directions. *Med Princ Pract.* 2015;24(suppl 1):14–28.
28. Li Y, Dong H, Li Y, Shi D. Graphene-based nanovehicles for photodynamic medical therapy. *Int J Nanomedicine.* 2015;10:2451–2459.
29. Matsushita-Ishiodori Y, Ohtsuki T. Photoinduced RNA interference. *Acc Chem Res.* 2012;45:1039–1047.
30. Boe S, Longva AS, Hovig E. Photochemically induced gene silencing using small interfering RNA molecules in combination with lipid carriers. *Oligonucleotides.* 2007;17:166–173.
31. Oliveira S, Fretz MM, Hogset A, Storm G, Schiffelers RM. Photochemical internalization enhances silencing of epidermal growth factor receptor through improved endosomal escape of siRNA. *Biochim Biophys Acta.* 2007;1768:1211–1217.
32. Boe S, Longva AS, Hovig E. Evaluation of various polyethylenimine formulations for light-controlled gene silencing using small interfering RNA molecules. *Oligonucleotides.* 2008;18:123–132.
33. Raemdonck K, Naeye B, Hogset A, Demeester J, De Smedt SC. Prolonged gene silencing by combining siRNA nanogels and photochemical internalization. *J Control Release.* 2010;145:281–288.
34. Bøe SL, Longva AS, Hovig E. A novel photosensitizer for light-controlled gene silencing. *Nucleic Acid Ther.* 2011;21:359–367.
35. Jin H, Lovell JF, Chen J, et al. Mechanistic insights into LDL nanoparticle-mediated siRNA delivery. *Bioconjug Chem.* 2012;23:33–41.
36. Dreaden EC, Mackey MA, Huang X, Kang B, El-Sayed MA. Beating cancer in multiple ways using nanogold. *Chem Soc Rev.* 2011;40:3391–3404.
37. Rosi NL, Giljohann DA, Thaxton CS, et al. Oligonucleotide-modified gold nanoparticles for intracellular gene regulation. *Science.* 2006;312:1027–1030.
38. Oishi M, Nakaogami J, Ishii T, Nagasaki Y. Smart PEGylated gold nanoparticles for the cytoplasmic delivery of siRNA to induce enhanced gene silencing. *Chem Lett.* 2006;35:1046–1047.
39. Giljohann DA, Seferos DS, Prigodich AE, Patel PC, Mirkin CA. Gene regulation with polyvalent siRNA-nanoparticle conjugates. *J Am Chem Soc.* 2009;131:2072–2073.
40. Hirsch LR, Stafford RJ, Bankson JA, et al. Nanoshell-mediated near-infrared thermal therapy of tumors under magnetic resonance guidance. *Proc Natl Acad Sci USA.* 2003;100:13549–13554.
41. Wijaya A, Schaffer SB, Pallares IG, Hamad-Schifferli K. Selective release of multiple DNA oligonucleotides from gold nanorods. *ACS Nano.* 2009;3:80–86.
42. Braun GB, Pallaoro A, Wu G, et al. Laser-activated gene silencing via gold nanoshell-siRNA conjugates. *ACS Nano.* 2009;3:2007–2015.
43. Chang YT, Liao PY, Sheu HS, et al. Near-infrared light-responsive intracellular drug and siRNA release using Au nanoensembles with oligonucleotide-capped silica shell. *Adv Mater.* 2012;24:3309–3314.
44. Huschka R, Barhoumi A, Liu Q, et al. Gene silencing by gold nanoshell-mediated delivery and laser-triggered release of antisense oligonucleotide and siRNA. *ACS Nano.* 2012;6:7681–7691.
45. Jeong EH, Ryu JH, Jeong H, et al. Efficient delivery of siRNAs by a photothermal approach using plant flavonoid-inspired gold nanoshells. *Chem Commun.* 2014;50:13388–13390.
46. Lu W, Zhang G, Zhang R, et al. Tumor site-specific silencing of NF-kappaB p65 by targeted hollow gold nanosphere-mediated photothermal transfection. *Cancer Res.* 2010;70:3177–3188.

47. Huang X, Pallaoro A, Braun GB, et al. Modular plasmonic nanocarriers for efficient and targeted delivery of cancer-therapeutic siRNA. *Nano Lett.* 2014;14:2046–2051.
48. Brown PK, Qureshi AT, Moll AN, Hayes DJ, Monroe WT. Silver nanoscale antisense drug delivery system for photoactivated gene silencing. *ACS Nano.* 2013;7:2948–2959.
49. Qureshi AT, Monroe WT, Dasa V, Gimble JM, Hayes DJ. miR-148b-nanoparticle conjugates for light mediated osteogenesis of human adipose stromal/stem cells. *Biomaterials.* 2013;34:7799–7810.
50. Yguerabide J, Yguerabide EE. Light-scattering submicroscopic particles as highly fluorescent analogs and their use as tracer labels in clinical and biological applications. *Anal Biochem.* 1998;262:137–156.
51. Yguerabide J, Yguerabide EE. Light-scattering submicroscopic particles as highly fluorescent analogs and their use as tracer labels in clinical and biological applications – II. Experimental characterization. *Anal Biochem.* 1998;262:157–176.
52. Schoolmeesters A, Eklund T, Leake D, et al. Functional profiling reveals critical role for miRNA in differentiation of human mesenchymal stem cells. *PLoS One.* 2009;4:e5605.
53. Zhang P, Steelant W, Kumar M, Scholfield M. Versatile photosensitizers for photodynamic therapy at infrared excitation. *J Am Chem Soc.* 2007;129:4526–4527.
54. Qian HS, Guo HC, Ho PC, Mahendran R, Zhang Y. Mesoporous-silica-coated up-conversion fluorescent nanoparticles for photodynamic therapy. *Small.* 2009;5:2285–2290.
55. Chen Y, Wang Y, Zhang J, et al. Rab10 and myosin-Va mediate insulin-stimulated GLUT4 storage vesicle translocation in adipocytes. *J Cell Biol.* 2012;198:545–560.
56. Jiang S, Zhang Y. Upconversion nanoparticle-based FRET system for study of siRNA in live cells. *Langmuir.* 2010;26:6689–6694.
57. Jayakumar MKG, Idris NM, Zhang Y. Remote activation of biomolecules in deep tissues using near-infrared-to-UV upconversion nanotransducers. *Proc Natl Acad Sci USA.* 2012;109:8483–8488.
58. Guo H, Yan D, Wei Y, et al. Inhibition of murine bladder cancer cell growth in vitro by photocontrollable siRNA based on upconversion fluorescent nanoparticles. *PLoS One.* 2014;9:e112713.
59. Yang Y, Liu F, Liu X, Xing B. NIR light controlled photorelease of siRNA and its targeted intracellular delivery based on upconversion nanoparticles. *Nanoscale.* 2013;5:231–238.
60. Jayakumar MKG, Bansal A, Huang K, et al. Near-infrared-light-based nano-platform boosts endosomal escape and controls gene knockdown in vivo. *ACS Nano.* 2014;8:4848–4858.
61. Wang X, Liu K, Yang G, et al. Near-infrared light triggered photodynamic therapy in combination with gene therapy using upconversion nanoparticles for effective cancer cell killing. *Nanoscale.* 2014;6:9198–9205.
62. Hou JQ, He J, Wen DG, Chen ZX, Zeng J. Survivin mRNA expression in urine as a biomarker for patients with transitional cell carcinoma of bladder. *Chin Med J.* 2006;119:1118–1120.
63. Al-Maghrebi M, Kehinde EO, Kapila K, Anim JT. Urinary survivin mRNA expression and urinary nuclear matrix protein 22 BladderChek and urine cytology in the detection of transitional cell carcinoma of the bladder. *Med Prin Pract.* 2012;21:295–297.
64. Endoh T, Sisido M, Ohtsuki T. Cellular siRNA delivery mediated by a cell-permeant RNA-binding protein and photoinduced RNA interference. *Bioconjug Chem.* 2008;19:1017–1024.
65. Endoh T, Ohtsuki T. Cellular siRNA delivery using TatU1A and photo-induced RNA interference. *Methods Mol Biol.* 2010;623:271–281.
66. Oubridge C, Ito N, Evans PR, Teo CH, Nagai K. Crystal structure at 1.92 A resolution of the RNA-binding domain of the U1A spliceosomal protein complexed with an RNA hairpin. *Nature.* 1994;372:432–438.

67. Matsushita-Ishiodori Y, Kuwabara R, Sakakoshi H, Endoh T, Ohtsuki T. Photosensitizing carrier proteins for photoinducible RNA interference. *Bioconjug Chem.* 2011;22:2222–2226.
68. Endoh T, Sisido M, Ohtsuki T. Spatial regulation of specific gene expression through photoactivation of RNAi. *J Control Release.* 2009;137:241–245.
69. Matsushita-Ishiodori Y, Morinaga M, Watanabe K, Ohtsuki T. Near-infrared light-directed RNAi using a photosensitive carrier molecule. *Bioconjug Chem.* 2013;24:1669–1673.
70. Park SJ, Park W, Na K. Photo-activatable ternary complex based on a multifunctional shielding material for targeted shRNA delivery in cancer treatment. *Biomaterials.* 2013;34:8991–8999.
71. Ma D, Lin QM, Zhang LM, Liang YY, Xue W. A star-shaped porphyrin-arginine functionalized poly(L-lysine) copolymer for photo-enhanced drug and gene co-delivery. *Biomaterials.* 2014;35:4357–4367.
72. Murayama S, Kos P, Miyata K, et al. Gene regulation by intracellular delivery and photodegradation of nanoparticles containing small interfering RNA. *Macromol Biosci.* 2014;14:626–631.
73. Weissleder R. A clearer vision for in vivo imaging. *Nat Biotechnol.* 2001;19:316–317.

CHAPTER FIVE

RNA Study Using DNA Nanotechnology

Hisashi Tadakuma[*,†,1], **Takeya Masubuchi**[†], **Takuya Ueda**[†]

[*] Institute for Integrated Cell-Material Sciences, Kyoto University, Kyoto, Japan
[†] Graduate School of Frontier Science, The University of Tokyo, Chiba, Japan
[1] Corresponding author: e-mail address: tadakuma.hisashi.8s@kyoto-u.ac.jp.

Contents

Abstract

Transcription is one of the fundamental steps of gene expression, where RNA polymerases (RNAPs) bind to their template genes and make RNAs. In addition to RNAP and the template gene, many molecules such as transcription factors are involved. The interaction and the effect of these factors depend on the geometry. Molecular layout of these factors, RNAP and gene is thus important. DNA nanotechnology is a promising technology that allows controlling of the molecular layout in the range of nanometer to micrometer scale with nanometer resolution; thus, it is expected to expand the RNA study beyond the current limit.

1. INTRODUCTION

Transcription is the first step of gene expression, where RNA polymerases (RNAPs) transcribe DNA to a corresponding RNA strand.

Progress in Molecular Biology and Translational Science, Volume 139
ISSN 1877-1173
http://dx.doi.org/10.1016/bs.pmbts.2015.11.004

The control of transcription, therefore, is the key of gene expression, hence the elucidation of the molecular mechanism of transcription has been one of the major issues of biology. In the past decades, many studies have identified RNAP and transcription-associated factors in detail. Structure studies by X-ray crystallography and electron microscopy revealed the precise molecular mechanism of RNAP, for example, phage T7 RNAP,[1] *Thermus thermophilus* RNAP,[2] and yeast RNA polymerase II,[3,4] at a single amino acid resolution. Recent advances in cryoelectron microscopy shed light on the spatial arrangement of RNAP and transcription-associated factors during the initiation of transcription. For example, 31-subunits assembly that contained a minimal yeast transcription initiation complex and the essential core of coactivator complex Mediator (15-subunits out of 25-subunits), was reported and the interaction between RNAP, general transcription factors, and Mediator was analyzed.[5] These structural data provide important insights into the geometry of the functional molecules yet they are basically static view. It would be difficult to fully describe the dynamic behavior of molecules solely by the structure study.

Single-molecule study is a powerful technique to observe the dynamic behavior of molecules, which is not hindered by averaging[6,7] (see chapters: Probing the Translation Dynamics of Ribosomes Using Zero-Mode Waveguides; −1 Programmed Ribosomal Frameshifting as a Force-Dependent Process of this book). As shown in chapter: Probing the Translation Dynamics of Ribosomes Using Zero-Mode Waveguides of this book, Joseph Puglisi and coworkers have been successfully applying single-molecule fluorescence imaging to ribosome study, and revealed many novel aspects of translation.[8–10] Similarly, dynamics of transcription by single-molecule *Escherichia coli* RNAP[11–15] and yeast RNA polymerase II[16] were reported using single-molecule fluorescence, magnetic trap, and optical trap microscopy. However, in these studies RNAP itself was not directly observed but small (eg, fluorescent dye) or large (eg, beads for magnetic and optical trap) markers attached to RNAP were. Another way to observe a single molecule is to use atomic force microscopy (AFM). AFM uses a cantilever with a sharp tip (probe) at its end that senses the difference in height. Using raster scan, the sample surface on mica plate was directly imaged and the dynamic behavior of proteins was visualized.[17–19] Nucleus, where transcription occurs, is not spatially homogeneous but rather heterogeneous. And the gene expression occurs in nanometer to micrometer-scale reaction field. For example, ribosomes are assembled in nucleolus, and transcription is thought to take place at discrete foci termed "transcription factories,"[20,21] where

4–30 RNA polymerase II molecules work within a mean diameter ranging from 40 to 130 nm.[22–28] In these reaction fields, factors are not freely encountered in the homogeneous space like in in vitro bulk reaction, but are associated with each other in an organized form on scaffold DNA, RNA, and protein.[29–31] Therefore the molecular layout of each factor in the reaction field is crucial. However, in the past, there was little way to control the molecular layout at nanometer resolution in the spatial range up to tens to hundreds nanometer. In contrast, recently developed DNA nanotechnology allows us to control the layout of molecules using the addressability of DNA nanostructure. Exploring the molecular mechanism using designed nano-reaction field is the emerging research field. In this chapter, we summarize some pioneer work of applying DNA nanotechnology on controlling enzymatic reaction.

2. DNA NANOTECHNOLOGY

In 1982, Seeman proposed to use DNA as a material for self-assembly of nanometer size structure.[32] DNA is an excellent building block because of (1) its predictable three-dimensional (3D) structure through Watson–Crick base pairing, (2) its addressability through chemical modification of DNA at a specific site, and (3) its physical stiffness in the context of use as a nanometer-size building block [persistence length of double-stranded DNA is 150 bp (50 nm)]. Seeman originally proposed to use "immobile" junction with sticky end, allowing sequence-specific association of sticky ends to create self-assembling arrays and self-assembled 3D lattices as scaffolds to orchestrate macromolecules into crystalline lattices (Fig. 1A). In other words, 3D DNA lattices could be used as hosts to organize guest-protein macromolecules and facilitate protein crystallography. More than 3 decades after this proposal, DNA nanotechnology is applied not only to crystallography, but also to vast fields of biology.

2.1 DNA-Based Nanostructure Construction

In homologous recombination, the Holliday junction is a key intermediate state, which contains four DNA strands forming four double-helical arms that flank a central branch point termed "crossover." There are three possible conformers of the Holliday junction, that is, an unstacked form and two stacked forms. In the absence of divalent cations such as Mg^{2+}, the unstacked

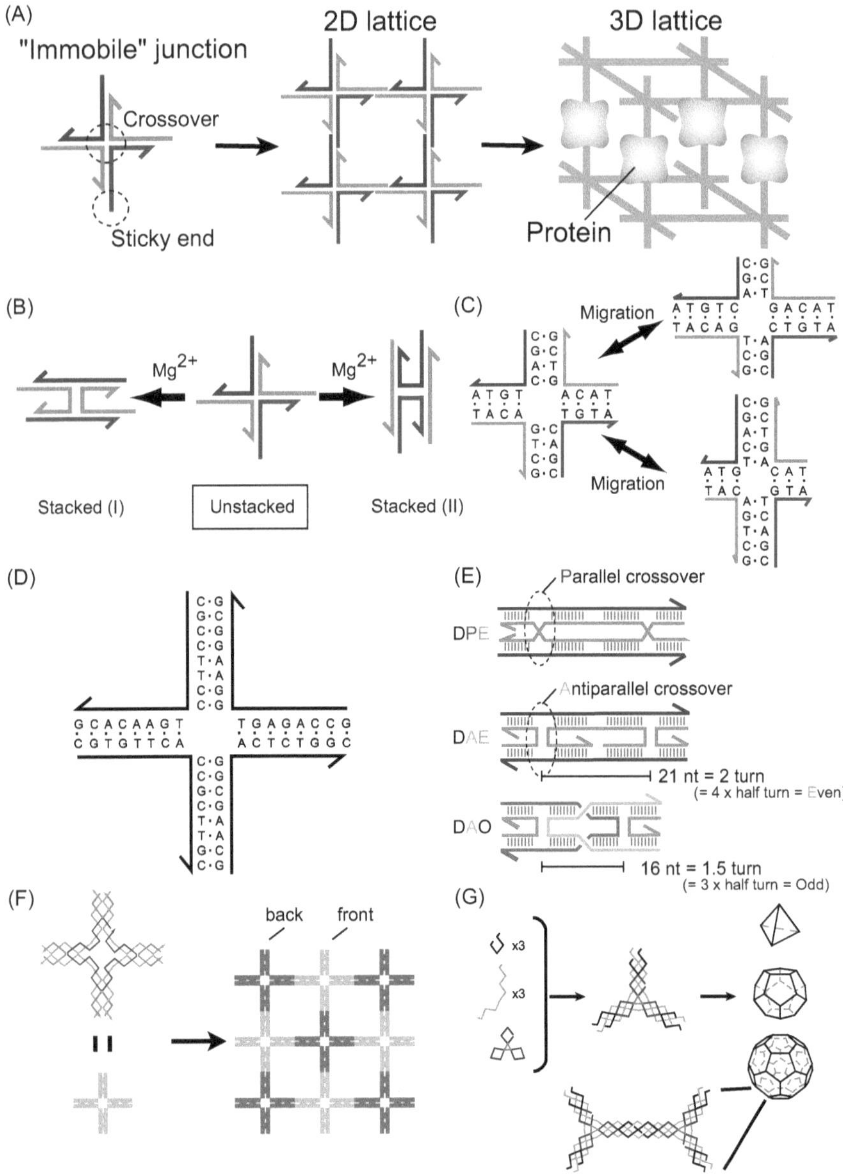

Figure 1 Multiple-stranded tile design. (A, left) Four single-strand DNAs form Holliday junction, which has a crossover (junction) and sticky ends, acts as a unit motif of higher-dimensional structures of 2D and 3D (right). The latter could be a scaffold to organize guest proteins to facilitate protein crystallization. (B) Under low ion concentration, native Holliday junction takes planar extended conformation "unstacked form" due to the electrostatic repulsion between the negatively charged backbones. It takes "stacked form" under high ion concentration. (C) Native Holliday junction also shows migration

form is dominant, because the electrostatic repulsion between the negatively charged backbones of the strands prevent the junction to take the stacked form, and therefore the unstacked form takes nearly planar extended conformation (Fig. 1B). In the presence of Mg^{2+}, the electrostatic repulsion is counteracted by Mg^{2+}, dominating the stacked structure, where two adjacent double-helical arms separated by an angle of about 60 degrees in a right-handed direction. With symmetrical sequences, the branch point "crossover" junction is mobile and shows branch migration (migrates in a random walk process). Therefore the overall structure of the Holliday junction is not unique but floppy (Fig. 1C).

To fix the unique structure, Seeman introduced two ideas, asymmetrical sequence and "double-crossover (DX)" molecule. With the designed asymmetrical sequence, base pairs are locked at the branch site, and the branch point is now "immobile" (Fig. 1D).[32,36] Further introduction of "double-crossover" molecule,[33] which has been proposed as an intermediate state in the double-strand break for recombination and contains juxtaposition of two crossovers between helical domains, the helices are stabilized in their parallel arrangement and held rigidly together. Based on these findings, motif-containing crossover has been used as a building block of DNA nanostructure.

There are several categories of DNA nanostructure design approach: (1) "multistrand tile design"[33,37] where multiple strands make a basic motif, (2) "single-stranded tile (SST) design"[38–41] where each single strand make a basic motif, and (3) "scaffold design"[42] which is composed of one long scaffold strand and multiple short helper strands. The last "scaffold design" is known as "DNA origami" because a single long strand is folded into a designed shape.

◀ (migrates in a random walk process). Therefore the overall structure of the Holliday junction is not unique but floppy. (D) An example of "immobile" branch point in Holliday junction obtained by means of sequence optimization. (E) In addition to "immobile" junction, "double-crossover (DX)" molecule was introduced to fix the unique structure. There are several types of DX molecule. (Top) DPE (double crossover, parallel, even number of half turns spacing), (middle) DAE (double crossover, antiparallel, even number of half turns spacing), (bottom) DAO (double crossover, antiparallel, odd number of half turns spacing), where DAE and DAO are only experimentally confirmed as stable structures. (F) Self-assembly of DX tile with periodic square cavities pattern. (G) Polyhedron and buckyball structures made by three-point-star motif method. *(Part A, C, D: adapted from Seeman (1982).[32] Part E: adapted from Fu and Seeman (1993).[33] Part F: adapted from Yan et al. (2003).[34] Part G: adapted from He et al. (2008).)[35]*

2.1.1 Multistranded Tile Design

"Multistrand tile design" is the originally provided approach based on "double-crossover (DX)" molecule. There are two types of DX molecule: parallel (P) and antiparallel (A) (Fig. 1E), expressing whether strands go forward or in the reverse direction at the crossovers, respectively. Depending on the number of helical half turns between crossovers, further subtype of even (E) and odd (O) exist. According to Seeman's acronym,[33] the distinct DX motifs could be described as DPE, DPO, DAE, DAO, where the first character D is for double crossover and the second character P or A indicates whether the structure is "parallel" or "antiparallel." In addition, for DPO, the extra half turn can correspond to a major-groove (wide) separation (W), or an extra minor-groove (narrow) separation (N), resulting two sub-subtypes DPOW and DPON. Among five distinct DX motifs, only two are experimentally confirmed as stable in small molecules with short arm: DAO (double crossover, antiparallel, odd number of half turns spacing) and DAE (double crossover, antiparallel, even number of half turns spacing). Therefore, these antiparallel forms of DX molecules have been used as the building block.

To make the 2D lattice, a single-strand sticky end with a unique sequence is attached at the corner of each DX molecule, so that the specific association of DX molecule is controlled.[37] Under rational design of sticky ends sequence of 5 to 6 base, each correct association results in the gain of 8–14 kcal mol^{-1} free energy at 25°C. The geometry of the DX lattice (termed DX-tile) is designed to resemble the natural twist of the B-form DNA double helix, in which a full turn takes place in 10.5 base pairs. For example, DAO molecules have an odd number of half turns between the crossover points (eg, 3 half turns of 16 base pairs), whereas DAE molecules have an even number of half turns (eg, 4 half turns of 21 base pairs). In addition to the arm length of the DX molecule, the length of sticky ends are optimized to make the separations between crossover points similar to that of B-form DNA, therefore adjacent DX molecules are coupled to keep coplanar, which was confirmed by AFM.[37]

As lattice is the programmable self-assembly product of motifs, the lattice pattern depends on the structure and sequence of the motifs. Using four 4-arm junction motifs, composed of nine strands of which one of the strands participating in all four junctions, periodic square-cavity pattern was observed (Fig. 1F).[34] In these patterns, corrugation strategy was used, where neighboring motifs are connected in alternating faces of an array plane to cancel potential curvatures associated with the DNA motifs (Fig. 1F). Similarly, self-assembly of six-point-stars shows lattice pattern with triangular and hexagonal pores.[43] But in this case, because of the limitation of rotational symmetry (sixfold rotational symmetry is not compatible with any twofold rotational symmetry lying in the plane of the lattice), corrugation strategy could not be applicable, and all motifs in a lattice have to face to the same side of the lattice. To satisfy this requirement, the separation between any two neighboring tiles has to be an integral number of turns (eg, 4 full turn of 42 base pairs).

3D structures of discrete polyhedral are also built up by multistrand tile design. First attempt to construct a closed polyhedral object from DNA was achieved by multistep ligation and purification of 10 oligonucleotide strands, making cube.[44] With similar approach, truncated octahedron was also made.[45] After a decade with more sophisticated sequence design technology, more complicated structures of tetrahedron[46] and buckyball (Fig. 1G)[35] were made by a simpler annealing method. In buckyball production, Mao and coworkers designed basic DNA-building units called three-point-star motif, which consists of three DX molecule arms linked via a central cyclic linker strand. Many copies of identical units then assembled into much larger 3D structures such as tetrahedron, dodecahedron, and buckyball with a diameter of 11, 24, and 41 nm, respectively. The destination structures depend on the flexibility and the concentration of the motif, suggesting that it might be possible to make more complex 3D objects with other motifs.

The first merit of this powerful design is that it requires only a small number of strands. For example, because of the rotational symmetry, the six-point-stars motif is composed of only three different single-strand species. Second, the lattice can expand to a wide area (eg, 20 μm). However, there are also demerits. The biggest drawback is that the multistrand tile design suffers from the difficulty of getting the strands ratio exactly equal. If the various component strands are not in equal proportions, incomplete structures would form. But multistrand tile design is still a good way to make a lattice on a surface.

2.1.2 Single-Stranded Tile Design

In contrast to multistrand tile design, where many copies of identical (or several types) motif units are assembled into much larger 2D or 3D structure, SST design is similar in concept to the rapid prototyping technology or LEGO block, each 3D voxel has a distinct sequence and combination of the block makes a structure. The strands have a floppy hairpin structure, and these strands are then assembled into the final structure by a cascade reaction. Therefore, the reaction could be proceeded isothermally, at a constant temperature (usually SST reaction is proceeded by normal annealing process). The floppy structure only folds into a block-like shape when incorporated into the assembly, which is in contrast to the multistranded tile design, where individual strands first fold into discrete building blocks and then individual blocks assemble into 2D or 3D structure. Due to this characteristic, in SST, reactions proceed via putative slow and sparse nucleation followed by fast growth, so that a large number of particles complete their formation not suffered by predominating partial structure formation occasionally observed in multistranded tile design.

Using 42-base SST strands. Yin and coworkers developed 2D nanostructures and simple 3D tube structure.[39,40] A similar approach was reported also by Mao and coworkers.[47] The SST strands have four domains with consecutive 10.5-base domains. In contrast to multistranded tile design, where the basic motif has a rigid structure "core" part and flexible "sticky ends," SST strands have only "sticky ends" (Fig. 2A). However, this floppy hairpin structure of sticky ends are folded upon incorporation into the structure, making a 3-by-7 nm tile attached to four neighboring tiles, which are acting as a pixel. After being folded, the SST strands have two 21-nucleotide antiparallel helices joined by a single phosphate linker (half crossover), which can be viewed as a simplified Holliday-junction analog. The crossover exchange point in SST is composed of single strand, rather than the normal double strands, so that it is termed "half" crossover.[47] Thermal formation and melting profiles of SST structures show hysteresis with a sharp transition temperature, suggesting that SST strands directly assembled into the growing lattice during annealing and disassembled from the lattice during melting. This character is in sharp contrast to that of multistranded tile design, where multiple characteristic transition temperatures were observed.[48,49] Therefore, the reaction could be driven isothermally, at a constant optimized temperature. Using hundreds of distinct SST strands, Yin and coworkers made more than 100 2D complex shapes (Fig. 2B), opening rapid prototyping technology at nanometer scale.

More complicated 3D structures are also made by Yin coworkers, termed "DNA bricks."[41] In one-pot reactions, hundreds of distinct sequences are self-assembled into desired target structures mediated by intermolecule-binding interactions. Here, each SST strand (32-base) consists of four consecutive 8-base domains (Fig. 2C), and has unique nucleotide sequence. Similar to the previously mentioned 42-base SST strands, the 32-base SST strands adopt an identical shape when incorporated into the target structure: two 16-nucleotide antiparallel helices joined by a single phosphate linker (half crossover). The two domains adjacent to phosphate linker (Domain 2 and 3 in Fig. 2C) are termed "head" domain, and the remaining two domains (Domain 1 and 4) are called "tail" domain. A DNA brick with a tail domain of sequence "a" (light-gray line in Fig. 2C) can interact with a DNA brick having a complementary "a★" head domain (dark-gray line in Fig. 2C). Each pairing between bricks defines three parallel helices packed to produce a 90-degree dihedral angle. This angle corresponds to approximately 3/4 helical twist of B-DNA attributed to the 8 base pairs. Each 8 base-pair interaction between bricks defines a voxel with dimensions of $2.5 \times 2.5 \times 2.7$ nm. With dimensions of $10 \times 10 \times 10$ voxels, a "molecular canvas" is defined, and selecting required subsets of bricks, more than 100 distinct shapes were designed. By changing the domain lengths of bricks, DNA bricks in hexagonal-, square-, and honeycomb-lattice structures can be made, where neighboring bricks form 60, 90, and 120 degree angles, respectively. Therefore, many 2D and 3D shapes could be prototyped using SST technology.

There are several merits of SST. First, in contrast to DNA origami technology described later in the chapter, where long scaffold-like M13mp18 phage single-strand circular genomic DNA of 7249 nucleotide is used, SST require only short strands. Therefore, in addition to natural DNA or RNA, many types of polymers, for example, DNA with chemically modified backbone or artificial bases, could be used. Furthermore, much different from multistranded tile design, where tight control of stoichiometry is the key, SST is more robust. Third, the variation of the SST strands is huge. For example, in the case of 32-nucelotide DNA bricks, the four domain bricks of 8 nucleotide could potentially reach 4^8 (524,288) varieties. This enables fabrication of nanostructures with tunable depths, and/or with complex 3D features (eg, channels or pores inside the structure).[50] Up till now the biggest structure reported is 80 nm for the depth and micrometer for the width,[50] but this would be improved in the future. Although promising, there are drawbacks of SST. Low yield of the product is the first one. Compared to multistranded tile design [eg, buckyball (69%)] and DNA origami technique (typically >90%), the reported yields of SST are between a few percent and 30%.[41] Better kinetic study and further improvement of sequence design technology are awaited. Second potential drawback is the weakness of the physical rigidity. In contrast to DNA origami, where long single-strand DNA scaffold and double-strands crossover supports the structural rigidity (see next section for detail), SST may be physically damaged from the weak interaction sequence. This point might be solved by cross-linking each strand after shape formation, or by using artificial nucleotide which has much stronger hybridization property. Despite these drawbacks, rapid prototyping of SST is advantageous and combining SST and DNA origami is a future challenge, where DNA origami's longer scaffold provides

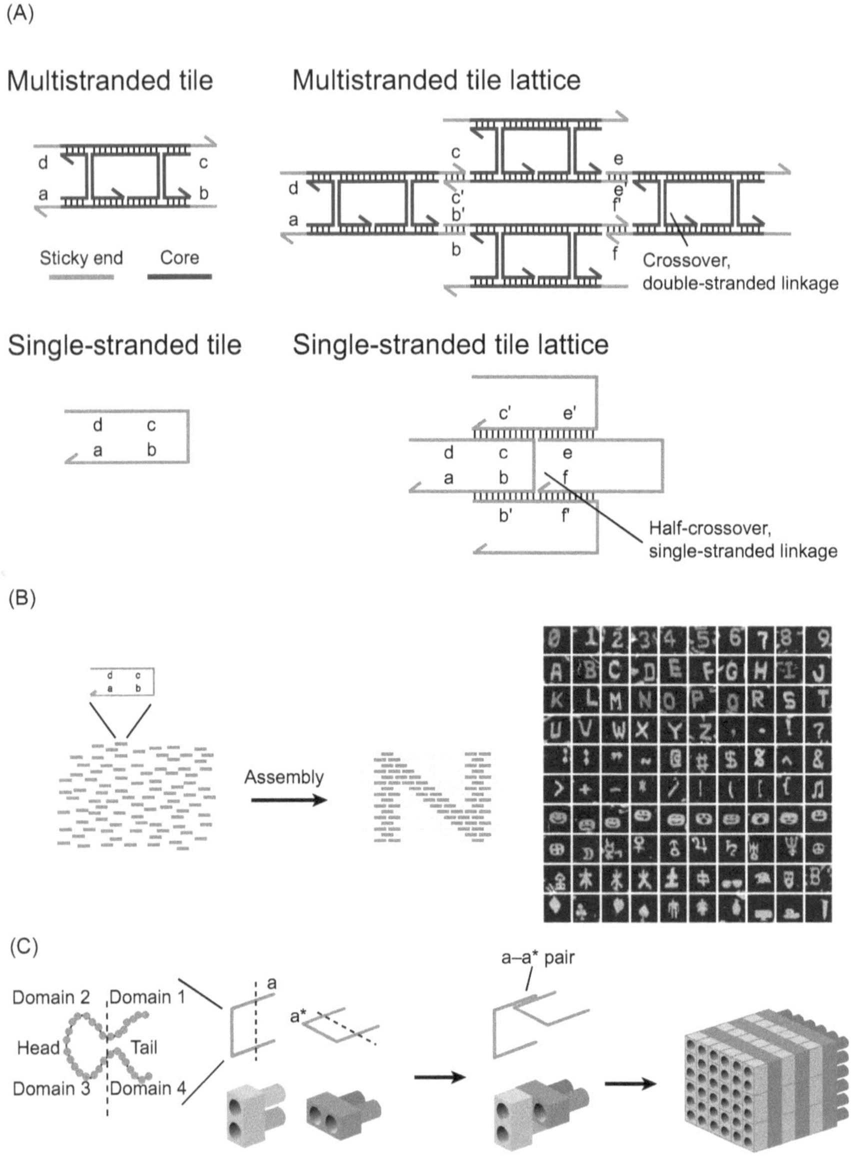

Figure 2 SST design. (A) One DNA strand acts as one motif in SST design, whereas four DNA strands are used in multistranded tile design. Also, crossover of SST is made of single-stranded linkage, while double-stranded linkage is in multistranded tile. (B) Over 100 2D structures were obtained in SST design. (C) Various 3D structures were designed and obtained by "DNA bricks" method, where the two domains adjacent to phosphate liker (Domain 2 and 3) are termed "head" domain, and the remaining two domains (Domain 1 and 4) are called "tail" domain. A DNA brick with a tail domain of sequence "a" (light-gray line) can interact with a DNA brick having a complementary "a*" head domain

better structural support, and SST's shorter strands may provide finer modularity and features.

2.1.3 Scaffold Design (DNA Origami)

DNA origami takes a much different approach from that of multistranded and SST design.[42] Usually, 7249-nucleotides long single-strand circular genomic DNA of M13mp18 phage is used as scaffold and hundreds of short helper strands called staples are used to fold longer scaffold into specific structure. In the folding process, scaffold DNA acts as a guide or a seed,[51] which increases the efficiency of folding and robustness to the stoichiometry of strands.

In 2006, "Smiley face" shook the DNA nanotechnology field (Fig. 3A).[42] Rothemund changed the rule of the game from assembling short strands motif into a large structure to fold long scaffold into specific structure.[52] The long scaffold of DNA origami is fold and hold by crossover made by staple strands. Typically, the staple strands bind to three adjacent helices, and the length is commonly 32 nucleotides, in which central 16 nucleotides bind to one helix and the remaining two parts of 8-nucleotide ends bind to the adjacent helices (Fig. 3B). The helical turn of DNA is usually approximated to be 3–3.5 nm in length and 3.5 nm in width. One helical turn (10.67 nucleotides) is different from that of canonical DNA (10.4 nucleotides/turn), resulting in the slightly twisted structure. Therefore, to relax the strain, usually one nucleotide is omitted every 48 nucleotides.[53] The length (about 3.5 nm) is also slightly different from that of canonical DNA (3.4 nm), which might be due to the interhelix gap presumably induced by electrostatic repulsion. Folded structures with straight edges sometimes stick together due to π–π stacking. To prevent this aggregation, single-strand 4T hairpin loops (four thymidines) are introduced to the staple strands located at the edge and corner part. If the stacking of folded DNA origami cause severe problem, one can design the edge with concavity and convexity.

Folding of DNA is performed by adding a 5- to 10-fold excess of each staple strand, and by annealing the sample using PCR machine with ramp method (decrease the temperature of the sample with time) or at constant temperature.[54] Folded DNA origami can be purified by column (ultrafiltration, gel filtration),

◀ (dark-gray line). Each pairing between bricks defines three parallel helices packed to produce a 90-degree dihedral angle. Each 8-base-pair interaction between bricks defines a voxel with dimensions of 2.5 × 2.5 × 2.7 nm. With dimensions of 10 × 10 × 10 voxels, a "molecular canvas" is defined. *(Part A: adapted from Liu et al. (2006)[47] and Yin et al. (2008b)[39]. Part B: adapted from Wei et al. (2012), images reproduced with permission from Nature Publishing Group (NPG)[40]. Part C: adapted from Ke et al. (2012).)[41]*

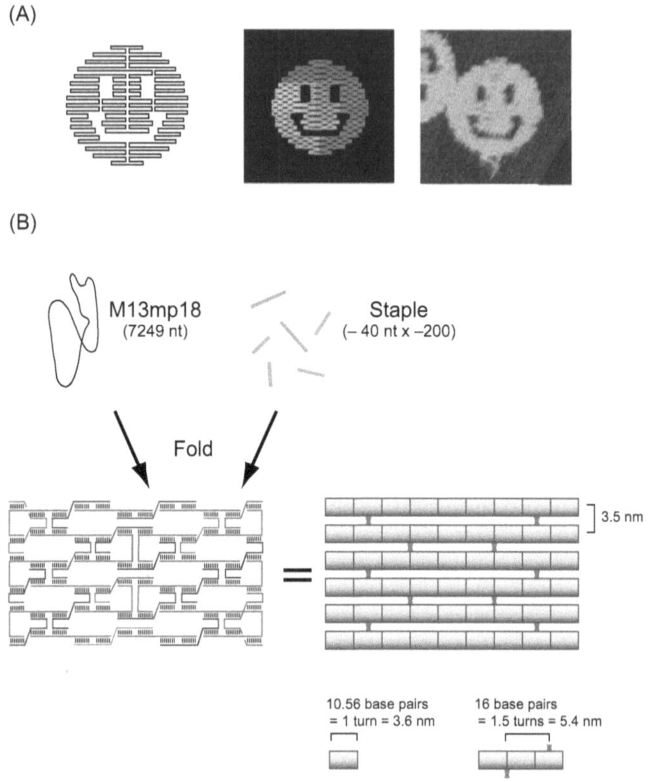

Figure 3 Scaffold design (DNA origami). (A) Smiley face structure with DNA origami method. (B) In DNA origami method, long circular single-stranded DNA (black) is folded into the desired shape by many short single-stranded DNAs (termed "Staple"), the latter typically bind to three adjacent helices, and the length is commonly 32 nucleotides, in which the central 16 nucleotides bind to one helix and the remaining two parts of 8-nucleotide ends bind to the adjacent helices. Unit pixel size is with dimensions of 3.6 × 3.5 nm. *(Part A, B: adapted from Rothemund (2006), images reproduced with permission from Nature Publishing Group (NPG).)*[42]

by gel electrophoresis,[55,56] or by PEG precipitation.[57,58] The yield of folding is quite high (90–95%) and the homogeneity is also high.[59]

DNA origami also can be folded into 3D objects,[60,61] where the architecture is developed from a six-helix bundle (6HB) DNA nanotube.[62] In this architecture the unit length is 7 nucleotides and not 8 nucleotides. Seven nucleotides correspond exactly to 2/3 of a turn and 14 nucleotides correspond exactly to 4/3 of a turn, therefore, crossover between adjacent helix is allowed in honeycomb 6HB bundle structure, where six helices rotated 120 degrees to each other (Fig. 4A). These features allow connecting multiple

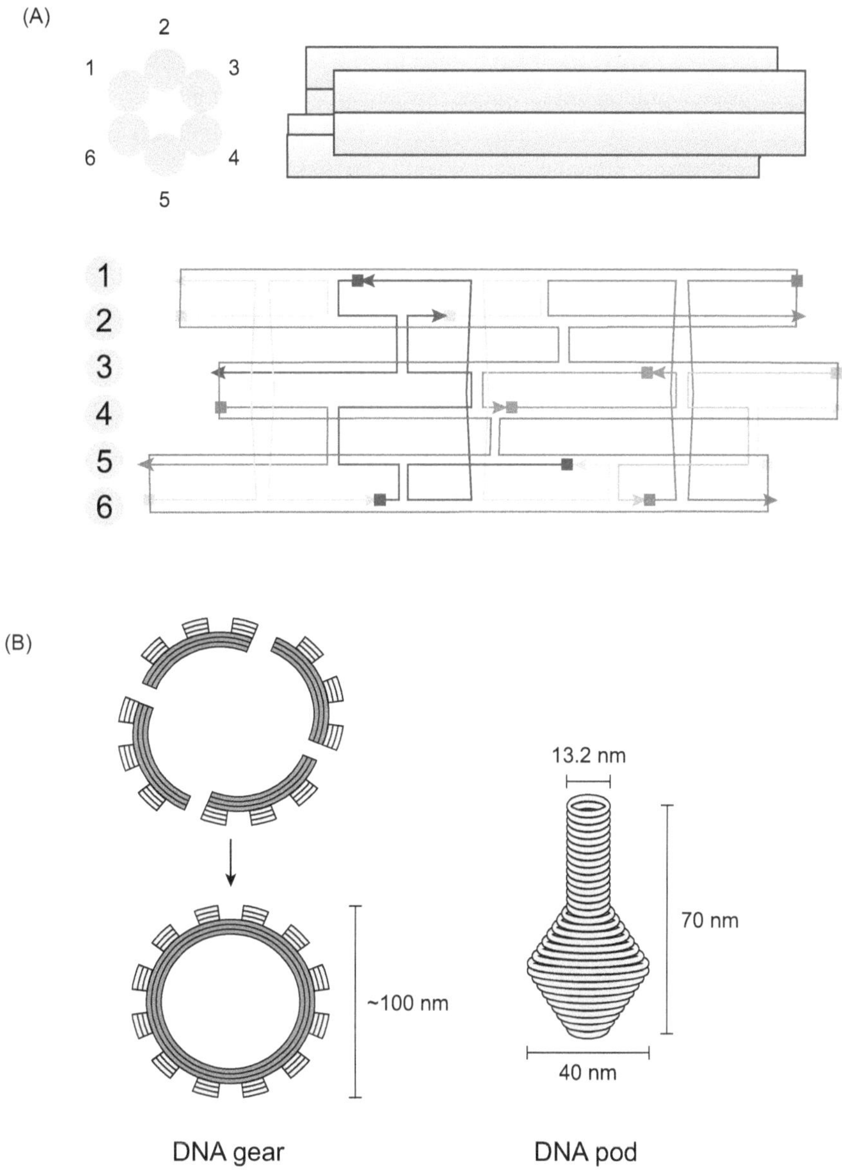

Figure 4 3D DNA structure of Scaffold design. (A) Basic unit of scaffold designed 3D DNA structure. Cross-section (left) and side view (right) of honeycomb structure composed of six helices with sample staples using caDNAno (http://cadnano.org/, bottom).[60,70] See also Section 5.3 for detailed design processes. (B) Representative 3D structures such as gear and pod. *(Part B: adapted from Deitz et al. (2009) (left)*[63] *and Han et al. (2011) (right).)*[67]

honeycomb layers, enabling the formation of 3D objects (Fig. 4A). Further introduction of twist and curve, more complicated structure of gear,[63] box,[64–66] pot,[67] and sphere[68] were made (Fig. 4B). Recently, with the aid of graph theory and relaxation simulation, a general method of folding arbitrary polygonal digital meshes into the structure was reported, in which the design process is highly automated.[69] Thus, various types of structures could be made by scaffold design (DNA origami) methods.

Highly assembled structure of DNA origami is also possible. Using pole and joint approach Yin and coworkers made hexagonal prism (60 MDa, Fig. 5A).[71] With 100-nm edges, the sizes of these structures become comparable to those of bacterial microcompartments such as carboxysomes. The joint pole termed DNA "tripod" is a 5-MDa 3-arm-junction origami tile, in which interarm angles and pole (arm) length can be controlled, so that, with the connector sequence design, many types of structures such as a tetrahedron (–20 MDa), a triangular prism (–30 MDa), a cube (–40 MDa), a pentagonal prism (–50 MDa), and a hexagonal prism (–60 MDa) can be self-assembled. In a tripod, each arm has an equal length (–50 nm) and contains 16 parallel double-helices packed on a honeycomb lattice with twofold rotational symmetry, and "struts" consisting of two double-helices support and control the angle between the two arms. The yields of tripod-assembled structures are highly dependent on the number of vertexes: 45, 24, 20, 4.2, and 0.11% for the tetrahedron, the triangular prism, the cube, the pentagonal prism, and the hexagonal prism, respectively. Dietz and coworkers took another approach to make polymerized structures with dynamic structure change.[72] Inspired by the interaction between an RNA-based enzyme ribonuclease P (RNase P) which cleaves the 5′ leader sequence for tRNA maturation and its substrate pretransfer RNA (tRNA). They used shape complementarity to assemble multiple DNA origami. In RNase P recognition, the acceptor stem and the TΨC loop of tRNA fit to the binding pocket of RNase P by a few nucleobase stacking interactions with the S domain of RNase P (Fig. 5B).[73] Similarly, in RNase P-inspired shape recognition method, blunt-ended double-helical DNA protrusions on one motif assume the role of the tRNA acceptor stem and corresponding concave on another motif mimic the RNase P binding pocket, and the nucleobase stacking bonds connect two motifs.[53,74–76] Upon two motifs engage, nucleobase stacking interactions occur at the double helical interface of the shape complementary protrusions and concave, but only when the helices fit correctly. Nucleobase stacking interaction method is sensitive to the concentration of counter ions, such as monovalent and divalent cations in

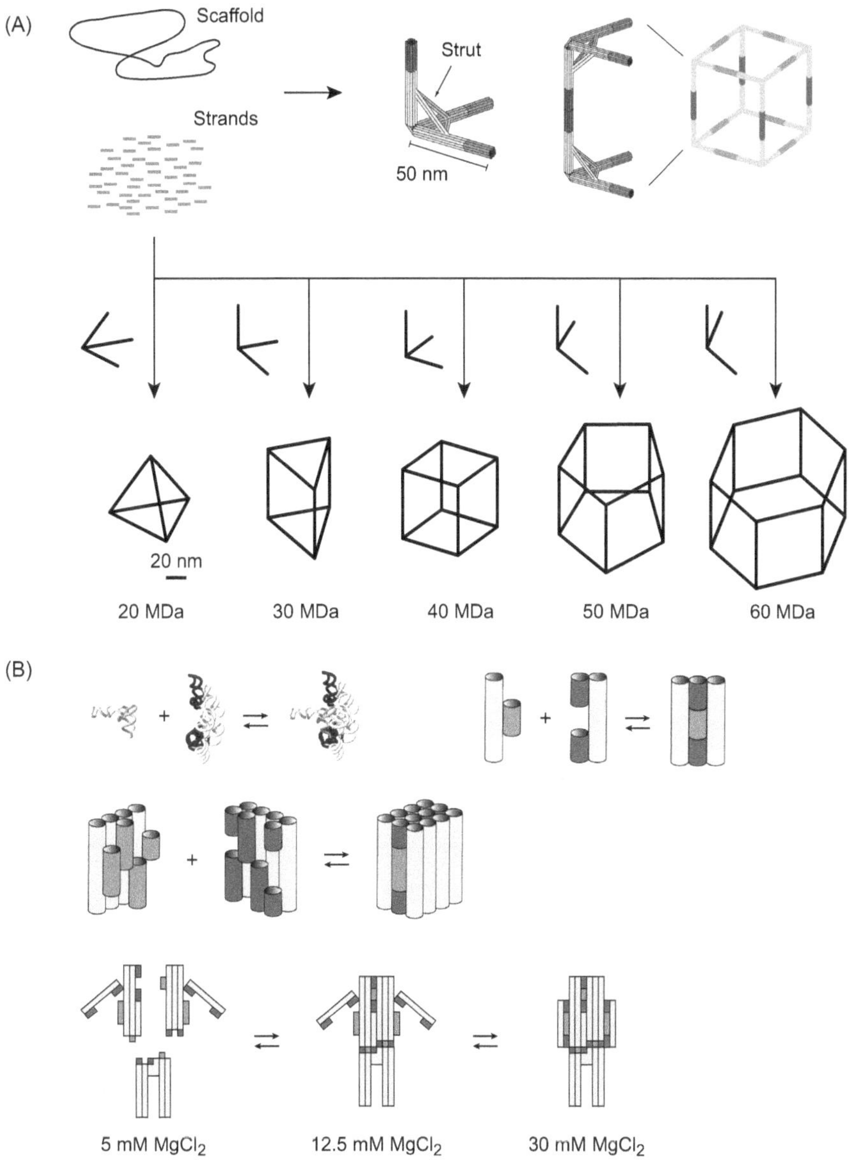

Figure 5 Higher assembled DNA structure. (A) Pole and joint approach to construct higher assembled DNA structure. A tripod is composed of one set of DNA origami structure and used as a basic unit that has three arms and binds with each other at the apical point. The structure of the end product is defined by the angle between the arms. (B) Shape-complementarity method to construct higher assembled DNA structure. (Top) DNA structure binds to its paired structure (top right) in a way that RNase P recognizes its substrate tRNA (top left). (Middle) Upon two motifs engagement,

the solution because repulsion between the negatively charged surfaces of DNA affects the equilibrium of the interaction, which allows on-off switching of the interface. All in all, without base pairing, RNase P-inspired method with nucleobase stacking bonds can build up micrometer-scale one- and two-stranded filaments and lattice, and transformable nanorobot. The merit of DNA origami is the design ability and robustness. As mentioned earlier, many types of structures have been made with high yield. In addition, long single-strand DNA scaffold can be a backbone, such that the structural rigidity to be ensured.[77] The limitation is based on the length of long scaffold. However, long scaffolds such as lambda DNA/M13 hybrid DNA scaffold (51,466 nucleotides),[78] PCR amplification-based scaffold [26 kb nucleotide fragment of lambda DNA (48,502 kb)],[79] and double strand form of lambda DNA itself[80] was used instead of M13mp18 scaffold (7,249 nucleotides). Combining these methods with higher order assemble method described previously, the limitation of scaffold may not be a problem from a practical point of view.

2.2 Functionalization of DNA Nanostructure

As described earlier, the method to fabricate DNA nanostructure has been much advanced. After structure fabrication, functionalization of fold DNA nanostructure is the key issue. Here we focus on the functionalization of DNA origami.

DNA origami has several advantages for organizing material: (1) predictable 3D structure through Watson–Crick base pairing, (2) addressability through chemical modification of DNA at specific site, and (3) physical stiffness in the context of use as a nanometer size breadboard (persistence length is 740 and 1880 nm for 4HB and 6HB, respectively).[81] Together, these advantages allow us to layout the molecules over long distance (10–1000 nm) while keeping nanometer resolution precision, which is much difficult using other techniques. There are two ways to address molecules: one is through noncovalent binding, and the other is through covalent binding.

◀ nucleobase stacking interactions occur at the double helical interface of the complementary shapes, but only upon correct fit of the helices. (Bottom) Shape-complementary interaction is highly affected by ion concentration, therefore higher assembled DNA structure, for example, nanorobot of 15 MDa, can be reversibly transformed in three different conformation states: disassembled, assembled with open arms, and assembled with closed arms, respectively, by changing the Mg^{2+} concentration. *(Part A: adapted from Iinuma et al. (2014).[71] Part B: adapted from Gerling et al. (2015).)[72]*

2.2.1 Functionalization Through Noncovalent Binding

There are several ways to couple materials to DNA origami through noncovalent binding. The most direct thus popular way is using the power of hybridization. After chemical conjugation of target materials to nucleic acid (eg, DNA, RNA, and artificial nucleic acid), material–nucleic acid complex is available for hybridization. Two typical ways for hybridization are (1) to use protrusion handle for hybridization and (2) to embed into the defect (hole) associated with the omission of specific strands at DNA origami production. In both cases, sequence of hybridization handle, temperature and concentration of divalent counter ion is crucial as is in usual hybridization. Because DNA is a rigid rod, structural hindrance may prevent the accessibility of immobilized material, especially for enzyme. However, in most cases, it does not cause a problem because single-strand DNA linker parts are introduced in some part between the DNA origami and anchored materials. Even if there is no such single-strand DNA linker part, there is actually a soft linker between DNA origami and the material: carbon linker of chemical moiety, typically 6–12 carbons, that connect DNA and materials. The flexibility of the carbon chains assures free rotations of the attached material around DNA because the single C—C bonds of the carbon linker act as swivels. Therefore, the effect of the direction of the tethered material does not need to be considered.[82,83] We confirmed this point using kinesin motor protein that walks along the microtubule.[84] We connected two motor domains (ATPase domains consist of 336–349 amino acids) with double-strand DNA (6–40 base pairs) to explore the precise mechanism of kinesin motility. With a proper double-strand DNA length, DNA–kinesin hybrid showed processive movement although the velocity and the run length (indicator of processivity) are about a half of those of natural protein-only kinesin, which might be improved if we can use fragile shorter double-strand DNA (<6 base pairs) experimentally. Later, Shih and coworkers succeeded in the attachment of two kinds of motor proteins (kinesin and dynein, the latter moves in the opposite direction to kinesin) through DNA hybridization method (Fig. 6A),[85] where they attached 1–7 motor proteins (occupancy yield is about 80%) and examined the effect of motor type, number, spacing, and orientation. In this case, natural protein-only motor attached to DNA origami through flexible linker showed normal behavior.

Avidin-biotin is a versatile method of attaching materials to DNA origami, as the dissociation constant (K_d) is in the order of 10^{-14} mol/L, which is one of the strongest noncovalent interactions known in nature, and avidin-biotin binding is robust to the condition (temperature, pH, detergents such

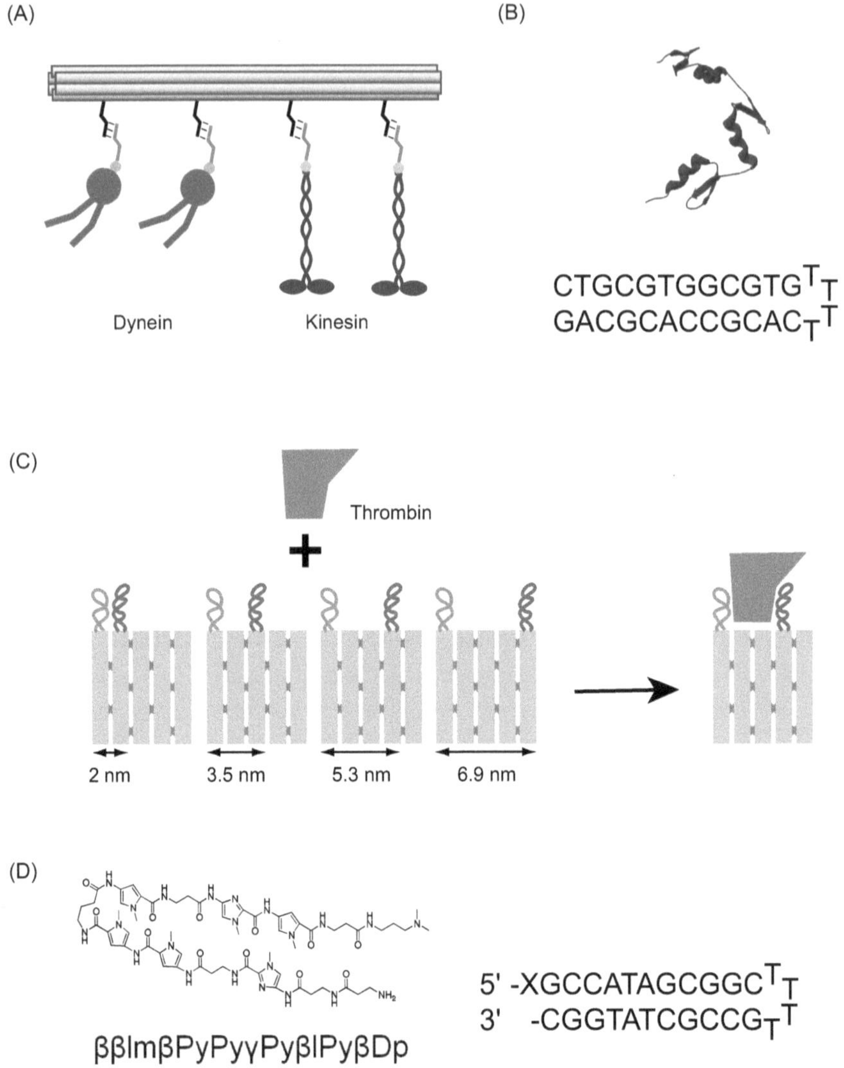

Figure 6 Functionalization of DNA nanostructure through noncovalent binding. (A) Hybridization method, where proteins, motor protein dynein, and kinesin in this figure, were organized through hybridization between the single-strand DNA attached to proteins and those embedded in the DNA nanostructure. (B) Zinc-finger proteins (ZFPs), that recognize 4 base pairs as one finger domain. Three-fingered protein recognizes 10 base pairs (bottom-light-gray characters) with nanomolar affinity. (C) To capture thrombin protein efficiently, Yan and coworkers used two types of aptamers (29 and 15 mer) that recognize different domains of thrombin and pinch thrombin in-between. The capture efficiency for 2, 3.5, 5.3, and 6.9 nm interaptamer distances were evaluated, revealing maximum at 5.3 nm with 40% occupancy yield and apparent kDa of 10 nM. (D) Pyrrole-imidazole polyamides (left) is a nucleotide-binding chemical product that can recognize specific-sequence colored light-gray (right). *(Part A: adapted from Derr et al. (2012).[85] Part B: adapted from Nakata et al. (2012).[86] Part C: adapted from Rinker et al. (2008).)[87]*

as SDS and organic solvents). Therefore, the application of avidin-biotin binding has been used from the early stage of DNA nanotechnology.[34] Gothelf and coworkers explored the occupancy yield using Rothemund's rectangle origami with 12 biotin positions,[88] and obtained 84% occupancy. Homotetramic streptavidin has four biotin-binding sites, which usually do not cause a problem, however if needed mono-, di-, and trivalent version of streptavidin is also available.[89,90] Furthermore, recent developments in avidin-biotin technology, such as modified biotin-binding pocket[91] or chemically modified biotin[92] allows switchable-binding control.

Fixing materials through nucleotide-binding protein is also a promising technique (Fig. 6B).[86] The merits of using nucleotide-binding proteins are (1) the sequence-specific binding allows orthogonal addressing, (2) fast-binding kinetics, and (3) (in case of protein anchoring) fusion of nucleotide-binding protein to the target protein allows omission of the modification process that is usually required before immobilization. Regarding the sequence specificity, nucleotide-binding protein usually recognizes 4–30 nucleotides, allowing a variety of sequence candidates albeit some mismatched sequence can also retain binding ability. The association constant (k_{on}) is quite high in nucleotide-binding protein especially in the case of RNA-guided nucleotide-binding proteins such as Cas-9 protein in CRISPR/Cas system and RNA-induced silencing complex (RISC) in RNA interference. In the case of RISC, which usually uses guide RNA to cleave target RNA, it is also known that bacterium *T. thermophilus* has a DNA-guided DNA silencing system, can bind to the target very fast (up to 4×10^8 M/s), that approaches to the limits of macromolecular diffusion (10^9 M/s), suggesting that RISC target binding is limited by the chance of two molecules' collision.[93] This is because the "seed" region of guide RNA is prearranged in an A-form-like helical structure by the partner Argonaute (Ago) protein.[94–99] Regarding the third merit, target protein fused to nucleotide-binding protein can easily bind to DNA nanostructure in one-pot reaction. Morii and coworkers explored one of the nucleotide-binding protein Zinc-finger proteins (ZFPs),[86] that recognizes 4 base pairs as one finger domain, and found that three-fingered protein recognizes 10 base pairs with nanomolar affinity and binds efficiently (45–75% occupancy yield) to the target sequence on the specific staple strands introduced into DNA nanostructure. However, the fusion protein with green fluorescent protein (GFP) variants only showed low efficiency of 30–50%, indicating the requirement of further improvement for practical use. Application of Cas-9 (or RISC) is also attractive,[100,101] but the huge size of the protein, *Streptococcus pyogenes* Cas9 has 1368 amino acids and recently reported smaller *Staphylococcus aureus* Cas9 still has 1053aa[102,103] or other CRISPR-related protein having single RNA guided Cpf1 has 1300aa,[104] might be a barrier for application. Nevertheless, the recent approach using guide sgRNA as a binding platform is attractive,[105,106] where extended guide sgRNA up to at least 4.8 kb is used as a protein-binding platform for selecting aptamers from the pool of RNA sequences.

Nucleotide base aptamer has also been used in DNA nanotechnology because of the straightforward incorporation ability of these aptamers into DNA nanostructure. Yan and coworkers explored the distance-dependent multivalent binding effects by incorporating multiple aptamers into DNA nanostructures (Fig. 6C).[87,107] To capture thrombin efficiently, they used two type of aptamers (29 and 15 mer) that recognize different domains of thrombin and pinch thrombin in-between, and evaluated the capture efficiency of 2, 3.5, 5.3, and 6.9 nm interaptamer distances, revealing maximum at 5.3 nm with 40% occupancy yield and apparent K_d of 10 nM. Although the variety of DNA aptamers is low compared to antibody, chemical modification of aptamer or artificial nucleotide aptamer may help to resolve the problem.[108]

Polypeptide-based binder, such as antibody is an attractive candidate.[109–112] Especially, recently developed small antibody (eg, nanobody[113]) or antibody-like protein (eg, affibody[114,115]) has advantages of enhanced stability and reduced size yet retaining similar target-binding characteristics of a full-size antibody (150 kDa). Nanobody is a small antibody with a single monomeric variable antibody domain with 12–15 kDa (about 110 amino acids), originally engineered from heavy-chain antibodies found in camelids.[113,116] Binding affinity is as strong as to subnanomolar range,[117] even after 2 h of incubation at 90°C,[118] which is mainly attributed to the efficient refolding activity of nanobody. Affibody is selected from phage displayed libraries having a single 58 amino-acid three helix bundle domain of staphylococcal protein A (SPA).[114,115] Although the small size (6 kDa), binding affinity is in the subnanomolar to picomolar range.[119] Taken together, polypeptide-based binder is attractive.

Another polypeptide-based binder is the purification tag such as His-tag.[120,121] Turberfield and coworkers examined the binding property of multivalent nitrilotriacetic acid (NTA) clusters.[121] They used DNA oligonucleotides with one, two, and three NTA groups (mono-, bis-, and trisNTA-functionalized oligonucleotides),[122] confirming the dissociation constant (K_d) between trisNTA and His-tagged protein (6 nM). In addition, similar to conventional column purification methods, addition of Ni^{2+} competitor (eg, imidazole) dissociated the linkage between trisNTA modified DNA and a His_6-tagged protein. Reversible binding and unbinding property of His-tag system may be useful for transient capture of molecules on DNA origami.

Nucleotide-binding chemical product (eg, pyrrole-imidazole polyamides) is an emerging method (Fig. 6D).[123,124] As described earlier, there are a variety of methods to immobilize molecules to DNA nanostructure, however, most of the methods require modification of DNA nanostructure prior to the immobilization process, such as introducing handle (eg, hybridization) or motif (eg, nucleotide-binding protein, aptamer), and conjugation of chemical moiety (eg, avidin-biotin system and His-tag system). In contrast, pyrrole (Py) -imidazole (Im) polyamides, which bind with the minor groove of DNA, can immobilize on naked DNA or histon-decorated DNA in a sequence-specific manner.[125] Py and Im recognize different bases. Py favors the T, A, and C but not G. In contrast, Im favors G. Thus, antiparallel pairs of Py/Im and Im/Py bind to C–G and G–C sequence, respectively, whereas antiparallel pair of Py/ Py binds to A–T and T–A. In addition, aliphatic b-alanine (b) can be substituted for Py, and antiparallel pair of Py/b and b/Py binds to A–T and T–A, while antiparallel pair of Im/b and b/Im binds to G–C and C–G. As the chemical synthesis of polyamides is easy and there are a variety of well-characterized sequences as well as chemical coupling methods, application of polyamides is attractive.

2.2.2 Functionalization Through Covalent Binding

In some cases, immobilization of target molecules through covalent binding is preferred. Here we explain several approaches.

Small tag protein that can bind orthogonal ligand is useful. Several types of such tag protein are available commercially (eg, Halo-tag and SNAP-tag). Halo-tag is developed from bacterial haloalkane dehalogenase to form a covalent bond with synthetic ligand.[126] Wild-type haloalkane dehalogenases remove halides from aliphatic hydrocarbons by a nucleophilic displacement mechanism. And a covalent bond is formed during catalysis between an aspartate in the enzyme and the hydrocarbon substrate. The mutation at an active site led to fix this intermediate, keeping a covalent bond between the Halo-tag protein and its ligand. The size of Halo-tag is about 33 kDa (297 amino acids, Fig. 7A) with the p*I* of 4.9. SNAP-tag has been developed by Johnsson and coworkers.[127] They developed SNAP-tag protein from the human DNA repair protein O-6-alkylguanine-DNA alkyltransferase (hAGT), which irreversibly transfers the alkyl group from its substrate, O-6-alkylguanine-DNA, to one of its cysteine residues. Given the substrate

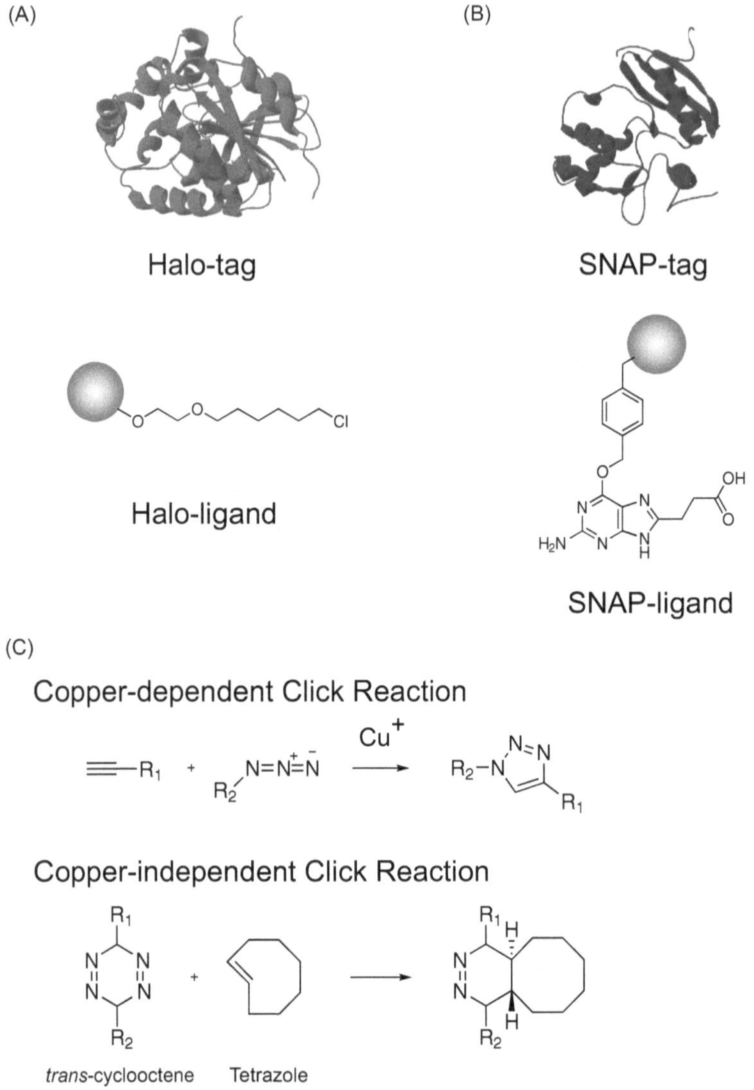

Figure 7 Functionalization of DNA nanostructure through covalent binding. (A) Halo-tag protein (297 amino acids, 33 kDa) and its substrate. (B) SNAP-tag protein (182 amino acids, 20 kDa) and its substrate. (C, top) Conventional click chemistry, where alkyne and azide react to form a triazole through Huisgen–Sharpless–Meldal copper(I)-catalyzed reaction, which is convenient but interferes with the formation of DNA nanostructure through potential Mg^{2+}–Cu^{2+} replacement and requires hazardous helper reagents for the reaction. (bottom) Cu-free click chemistry, where azide reacts with strained alkynes, such as cyclooctenes. Slow reaction speed (10–100 M/s) is the drawback of Cu-free click reaction.

specificity of hAGT is relatively low and similar O-6-benzylguanine (BG) is also accepted as a substrate, they optimized the substrate and did mutagenesis to obtain SNAP-tag (Fig. 7B). SNAP-tag is a slightly smaller 20 kDa protein (182 amino acids) with the p*I* of 6.0. Niemeyer and coworkers explored the binding ability of Halo-tag and SNAP-tag to DNA origami.[128] They found that the p*I* of protein affects the occupancy yield: on average 35 and 60% for Halo-tag fusion protein and SNAP-tag fusion protein, respectively. Using SNAP-tag, direct immobilization of dynein motor proteins onto DNA origami was reported.[58]

Click chemistry is a convenient method to conjugate molecules. In click reaction, alkyne and azide react to form a triazole through Huisgen–Sharpless–Meldal copper(I)-catalyzed reaction (Fig. 7C).[129,130] However, the high concentration of Cu^{2+} ion may replace the Mg^{2+} ion within DNA and the copper(I)-stabilizing ligand THTA (tris-(1-[3-hydroxypropyl] triazolyl-4-methyl)amine) interferes with the folding of DNA origami.[131] Nevertheless, click reaction on mica-tethered DNA origami was observed.[132] Recently developed Cu-free click chemistry resolved the drawback, where azide reacts with strained alkynes, such as cyclooctynes. Several types of Cu-free reaction were reported,[133] for example, a 1,2,4,5-tetrazine (Tz) with a *trans*-cyclooctene (TCO, Fig. 7C).[134–136] The main drawback of Cu-free reaction is its slow reaction speed (10–100 /M/s). Further chemical study may overcome this point.

3. HOW TO MAKE DNA NANOSTRUCTURE

The merit of DNA nanostructure is the predictability of the 3D structure from the 1D sequence information. This is in sharp contrast to the case of protein, where 3D structure prediction is still a challenge. Here we describe how to make DNA nanostructure, especially DNA origami.

There are mainly four steps:

1. design of staples
2. mix and fold
3. purification
4. observation or assay

The first thing to be done is to decide the design. There are several programs that assist the designing. At this moment (2015), the de facto standard one is the caDNAno (http://cadnano.org/), an open-source computer-aided design (CAD) software package with a graphical user interface, developed by William Shih and Shown Douglass' group.[70] The caDNAno works as a standalone software or a plugin for Maya (Autodesk), in the latter mode Maya draws a 3D model of the designed object simultaneously. The design process is divided into four steps. First, the outer shape of the target is approximated by a scaffold drawn by raster-style path (Fig. 8A), in which possible points of crossover are indicated and can be selected. Second, staple paths complementary to the scaffold are automatically assigned by "AutoStaple" function. By the default setting, all permitted staple crossovers are used, except for those that would be five base pairs away from a scaffold crossover between the same two helices. Third, using the built-in function "AutoBreak," the predicted staple paths are automatically broken into shorter segments of user-defined length range. After that each staple can be manually optimized for the length and the position (Fig. 8B). Finally, the sequence is defined with the scaffold (selected from 7 types ranging from 7249 to 8634 nucleotides). For estimating thermal stability of the designed structure, CanDo is useful (http://cando-dna-origami.org/).

Normal desalting-grade oligonucleotides are enough for staples. In the case of SST, it might be better to choose good quality oligos. Usually small-scale synthesis is enough. Regarding scaffold, several companies provide the M13mp18 single-strand DNA with a lot of stock.

Preparation of staple mixture is somewhat troublesome and it is recommended to make a series of premix containing 5–10 staples. Otherwise, machine-aided protocol is also available.[40] The ratio of staples to scaffold is typically 5–10 times, in which the scaffold concentration is 5–50 nM depending on the structure and buffer condition (typically 10–20 mM Mg^{2+} with 5–20 mM Tris-HCl (pH 7.5–8.3), 1 mM EDTA and 0–5 mM NaCl). The concentration of Mg^{2+} should be optimized to maximize the folding efficiency and to minimize the aggregation of DNA origami. Addition of monovalent ion such as NaCl, up to 5 mM, may help to reduce the aggregation. For folding, there are two ways: ramp method and constant temperature method. Ramp method is a conventional way of folding, decreasing the temperature with time. The slope for decreasing the temperature highly depends on the structure. For example, single-layered Rothemund's rectangle tile can be folded within 30 min by starting from 85°C to 25°C, lowering 1°C every 30 s. In contrast, complicated 3D structures take a few days to a week for folding.[61] Constant temperature method relies on the fact that DNA origami has a unique folding and melting temperature.[54] With a constant temperature method, Rothemund's rectangle tile is folded within 5 min and complicated structures are folded within 4–5 h. The critical temperature, typically around

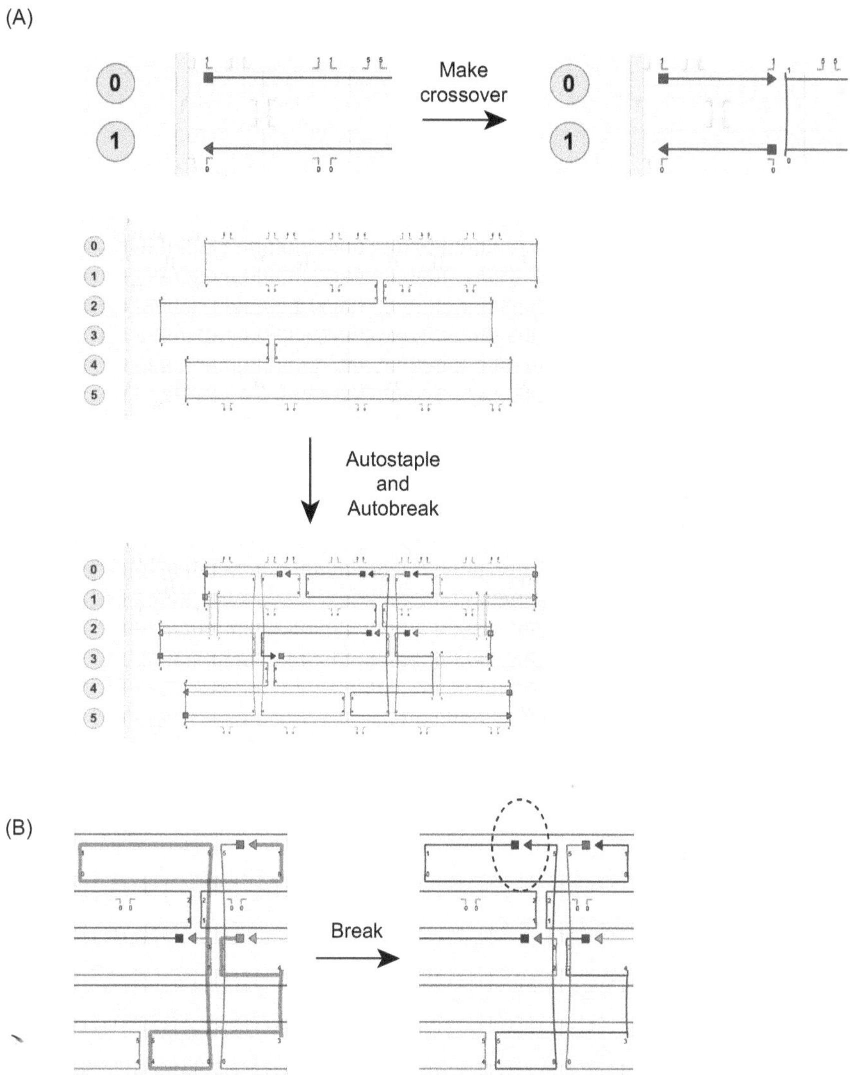

Figure 8 Design software assisted DNA nanostructure design. (A) Captured images of caDNAno (http://cadnano.org/), an open-source computer-aided design (CAD) software package with a graphical user interface, developed by William Shih and Shown Douglass' group.[70] By clicking the mark on the vicinity of the line, crossovers can be introduced at the predicted potential sites. After drawing the scaffold design, "Autostaple" and "Autobreak" functions fulfill the suggested staples. (B) With manual adjustment, final staple sequences are determined.

50°C, depends on the structure and the staple sequences. Therefore, pilot study should be performed. We usually try several temperature points between 40 and 60°C, typically 45–55°C, for 4 h.

Excess staples are removed by several ways. The most solid way is by agarose gel electrophoresis, although the method is time consuming and laboring: typically 1% gel in 0.5× TBE (or TAE) + 11 mM $MgCl_2$ with 50 V for 2–4 h. It should be noted that agarose gels and the running buffer should contain magnesium. However, high Mg^{2+} concentration causes heat production that disturbs electrophoresis and also melts the DNA origami structure. To avoid this melting of origami structure during electrophoresis, the migration should be done in the cold room, in refrigerator or with cooling system. The DNA origami appears as a band that migrates fastest. The band can be cut out from the gel and put into the squeeze spin column (eg, Biorad cat#732-6165), followed by crashing with plastic pistil and spinning down for the recovery. The procedure usually results in 2–5 nM DNA origami solution using 50 nM scaffolds as starting materials. To concentrate the solution, it is convenient to precipitate with 7.5% PEG-8000 (w/v)[57,58] [mixing sample 1:1 with "precipitation buffer" (15% PEG 8000 (w/v), 5 mM Tris-HCl (pH 7.5–8.3), 1 mM EDTA, and 500 mM NaCl, while keeping the final Mg^{2+} concentration higher than 10 mM] and centrifuge at 16,000 G for 5–25 min at room temperature). PEG precipitation alone is also useful to remove the excess staples, but in this case, aggregated DNA origami and unfolded scaffold are also precipitated. Another way to remove the excess staples is to use a spin column: ultrafiltration and gel filtration, 100-kDa Amicon filters, (cat#UFC510024 Millipore), or MicroSpin S-300 HR Columns, (cat#27-5130-01 GE, also S-200, S-400, and S-500 are available). Similar to PEG precipitation, aggregated DNA origami and unfolded scaffold cannot be removed with these methods.

Finally the obtained structures were observed by electron microscopy, for example, negative stain with Uranyl-Formate or Uranyl acetate,[55] or fluorescence microscopy including super-resolution microscopy.[71]

4. APPLICATION OF DNA NANOTECHNOLOGY IN RNA BIOLOGY

DNA nanostructure allows to place molecules in the range between nanometer to micrometer field with addressing resolution at nanometer level. Using these unique tools, a new way of studying RNA biology is opened.

4.1 Imaging of RNA Using DNA Nanostructure

"Seeing is believing" Richard Feynman's famous quote is also meaningful in RNA biology. As shown in chapter: Probing the Translation Dynamics of Ribosomes Using Zero-Mode Waveguides, Puglisi and coworkers unveiled important mechanisms of translation using single-molecule fluorescence imaging.[8–10] Recently, we applied single-molecule technique to explore the molecular mechanism of RNA interference,[98,137] in which we precisely analyzed the intermediate state and defined the fundamental step of RNAi effecter RISC (RNA-induced silencing complex) assembly[137] and also revealed the molecular mechanism how RISC balance the efficiency and accuracy.[98] In these studies, the key is how to discrete the meaningful signal from background noise signal. Confinement of the detectable area is the key idea. Given the constant signal intensity of the target molecule, and given the constant background signal per unit volume, reducing the observation volume results in the improvement of signal to noise ratio. After Betzig and coworkers opened the window of single fluorescence molecule study by observing single dye on the dried surface using near-field scanning optical microscopy (SNOM),[138] Yanagida and coworkers boosted the single-molecule biology by observing the single enzyme reaction in the solution using total internal reflection microscopy (TIRF, Fig. 9A).[6] Both SNOM and TIRF limit the excitation volume by small aperture or by evanescent field. The TIRF method is simple yet powerful, encouraging many biologists to explore their biological issues using single-molecule fluorescence imaging. But the excitation depth of TIRF, hundreds of nanometers, is still bulky, so that good signal noise ratio is obtained only at low fluorescent-dye concentration, typically up to 50 nM. After years, Webb and coworkers developed zero-mode waveguide (ZMW).[139] Using the top-down nanofabrication approach they made a tiny hole in the metal layer, typically 100 nm, coated on the glass surface. Using ZMW, the reduction of the excitation volume is achieved not only by the confinement of the depth direction, where vertical z direction excitation depth is reduced 10 times from that of TIRF (300 nm) to 30 nm (Fig. 9A), but also by the confinement of x–y horizontal direction (–100 nm), achieving zeptoliter sample volumes and hence the single-molecule fluorescence imaging in the presence of micromolar background fluorescent dye.

In addition to such "Top down" fabrication approach, where starting materials are cut into the final form, "Bottom up" fabrication approach such as DNA nanotechnology, where starting materials are self-assembled into the final form, also allows us to design the molecular layout at nanometer

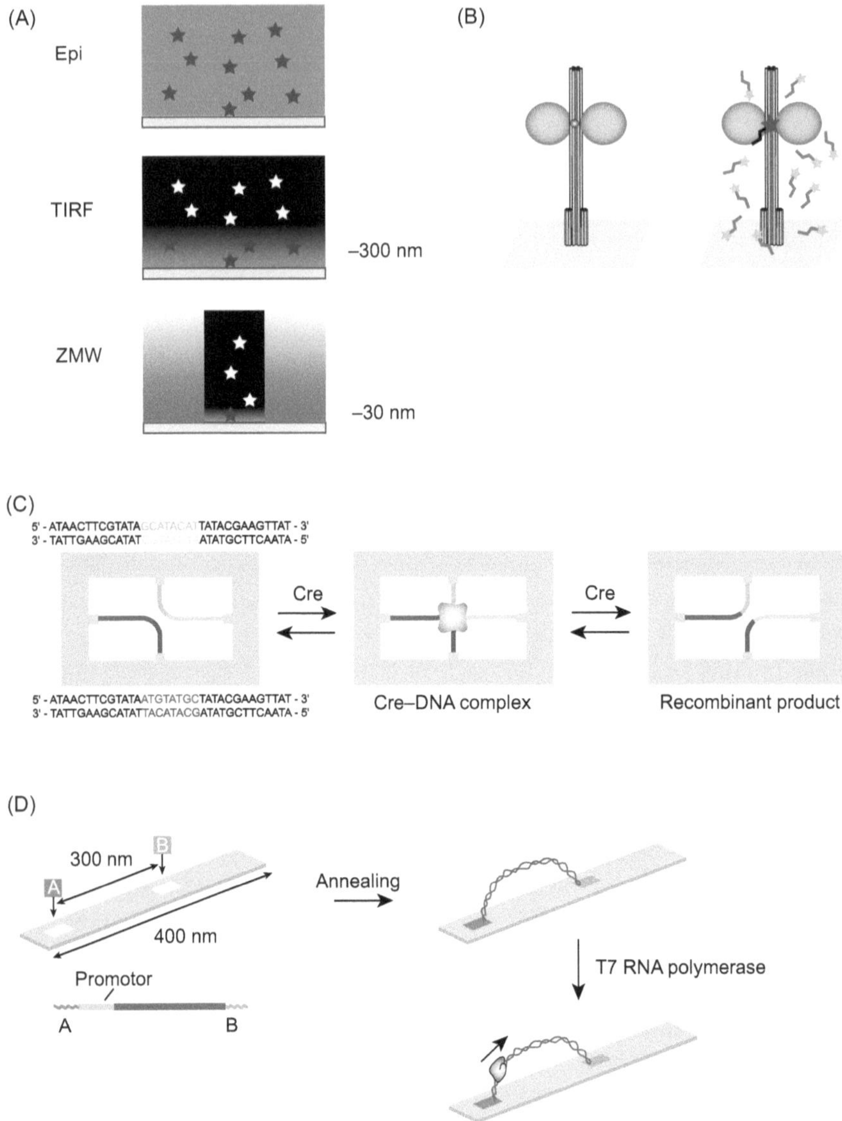

Figure 9 Imaging of RNA using DNA nanostructures. (A) Confinement of the excitation volume is the key for single-molecule imaging with high signal to noise ratio. Epi: epi-fluorescence imaging, where all fluorescent dyes are excited, making high background noise in observing single molecule anchored on the glass surface (bottom thin block). TIRF: total internal reflection fluorescence microscopy reduces the excitation volume in the depth direction, illuminating vertical (*z*) direction of up to 300 nm, reducing the background noise and enabling to observe single enzyme reaction in solution.[6] ZMW: in addition to *z*-direction, *x*–*y* horizontal direction is also confined (–100 nm), achieving zeptoliter sampling volumes and single-molecule imaging under the presence of

resolution such that target molecules are highlighted. The confinement of the excitation volume in between two gold nanoparticles, termed nanoantenna, allows us the zeptoliter sample volumes.[140] DNA origami also allows us to confine the reaction at the specific place, making the observation much easier.[143] The details are described later in the chapter.

4.1.1 Nanostructure Base Fluorescence Imaging

The ZMW method is powerful, however, the key ZMW chip is much expensive to fabricate, requiring sophisticated fabrication procedures performed in a clean room facility. Therefore, only limited people can afford to execute the experiments. In contrast, bottom-up approach such as DNA origami is much easier to fabricate, requiring just normal equipments for biochemists and molecular biologists, such as PCR machine. Tinnefeld and coworkers attached two 100 nm gold nanoparticles on DNA origami that also has docking sites for a single fluorescent dye in the 23-nm gap between two gold nanoparticles, enhancing the fluorescence intensity in a plasmonic hotspot of zeptoliter sample volume (Fig. 9B).[140] A pillar-shaped DNA origami has a length of 220 nm, a 15-nm diameter rod consisting of a 12HB and three extra 6HB on the base. After anchoring the pillar-shaped DNA origami on the glass surface, two DNA-functionalized gold nanoparticles were attached to the handle extruded from the pillar through DNA hybridization. In-between the two gold nanoparticles, enhanced local plasmonic fields are created such that fluorescent intensity enhancement occurred due to the quantum yield enhancement and the normalized electric field intensity enhancement. With this nanoantenna, binding and unbinding of single-strand DNA to the hotspot in-between two gold nanoparticles was observed at the concentration of 100 nM. In addition, fluctuation of a Holliday junction was also observed by fluorescence resonance energy transfer (FRET) at single-molecule level. Concentration of the fluorescent probes has to be limited to 100 nM for single-molecule detection, this is much lower

---

micromolar background fluorescent dye.[139] For details see chapter: Probing the Translation Dynamics of Ribosomes Using Zero-Mode Waveguides written by Tsai et al. (B) Nanoantenna has two 100 nm gold nanoparticles on a pillar-shaped DNA origami of 220 nm height, achieving single-molecule imaging in-between the gold particles. (C) Real time visualization of Cre recombinase reaction on DNA origami achieved by high-speed atomic force microscopy (HS-AFM). (D) Sliding of T7-RNA polymerase (T7-RNAP) on a template gene embedded into DNA origami was also visualized by HS-AFM. *(Part B: adapted from Acuna et al. (2012).[140] Part C: adapted from Suzuki et al. (2014).[141] Part D: adapted from Endo et al. (2012).)[142]*

compared to ZMW, where single-molecule imaging can be achieved using micromolar dye concentration. Nevertheless, optimized nanoantenna combining with the precise molecular layout character of DNA origami might be a powerful tool to explore the precise mechanism of RNA biology.

DNA origami can also be combined with ZMW. Despite the small sample volume of zeptoliter the diameter of ZMW (100 nm) is still quite large compared to the size of a single protein (5 nm). Furthermore the metal (aluminum or gold) wall of ZMW affects the dye properties. Therefore, a technique to centralize the molecule inside the ZMW hole might be useful. Tinnefeld and coworkers addressed this point by anchoring a DNA origami disc at the bottom of the ZMW chip.[144] Using this method, they investigated the quantum yield and excitation rate of single molecules that locate the vicinity of the ZMW metallic wall.[145] This approach of ZMW origami would be developed further.

4.1.2 AFM Imaging of DNA Origami Base Nanosystem

HS-AFM is a tool to visualize directly the real movement of a molecule in the solution, which is developed by Ando's and Hansma's group independently.[7,146] With the recent progress contributed by Ando's group, the movement and the activity of enzyme were captured in real time.[17–19] Therefore the application of HS-AFM into the RNA biology is an attractive issue. However, the rail (ie, DNA) on which motor proteins such as RNA polymerase walks along is much softer than the ones previously reported. For example, the persistence length of DNA is about 50 nm.[147] This is much smaller than myosin's rail, actin filament (17.7 μm), so that it requires some solid support to visualize. Furthermore, the molecular layout of the associated factors is important in the reaction, therefore, molecule layout technology with nanometer resolution is required. DNA origami is the technique that fulfills both these requirements. Sugiyama and Endo's group have performed some pioneering works.[148]

They created frame structure, where a window inside the origami structure serves as a reaction field to monitor the movements of molecules (Fig. 9C). For example, they embedded DNA substrate strands of a restriction enzyme[149] and Cre recombinase.[141] In the case of the restriction enzyme, a structurally tensed 64-mer strand and a relaxed 74-mer strand were incorporated directly into the DNA origami frame. The activity of the M.EcoRI modifying enzyme was studied that introduces a methyl group at the second adenine of the GGAATC sequence in the presence of S-adenosyl-l-methionine (SAM). In the methylation process, M.EcoRI bends the double helix, therefore, the relaxed strand that can be bended served as a better substrate compared to the tensed strand. This was evidenced by subsequent treatment with the restriction enzyme R.EcoRI after the methylation by M.EcoRI. Using HS-AFM they observed the cleavage of the relaxed strand, whereas the tensed strand was not efficiently cleaved. Similarly, the recombination process by Cre recombinase was visualized, showing the potential of HS-AFM in gene expression related biological processes (Fig. 9C).

Transcription process was also visualized with HS-AFM and DNA origami.[142] This time the rail DNA, 1000-base-pair DNA, is incorporated into the DNA origami structure of 400 × 20 nm, and the sliding movement of T7 RNA polymerase was observed. The two ends of template DNA are anchored at specific positions such that the transcription direction is predictable. Furthermore, to distinguish the template DNA orientation by HS-AFM, the two positions that anchor template DNA were designed to locate at different distances from the edge of the DNA origami platform. The RNAPs that slide along the template DNA were distinguished from nonspecifically bound RNAPs on the mica surface by two criteria: (1) whether the sliding is along the template DNA, and (2) whether the height is higher than that of RNAPs alone. Upon addition of RNAPs onto the DNA origami platform, the association onto the promoter region, the sliding and the dissociation from the template DNA of RNAPs were observed. Furthermore, the transcribed nascent RNAs were visualized using biotinylated UTP and streptavidin decoration. The nascent RNAs were observed under the existence of NTPs, but not under the condition omitting UTP, indicating the success of imaging. However, as the association frequency is quite low, this assay is laborious and difficult to apply for further investigation of transcription mechanism. To overcome this point, simultaneous anchoring of the RNAPs and template DNAs on the DNA origami may be a promising approach, as in such an integrated system would facilitate RNAPs to encounter their substrates, template DNAs. We investigated RNA transcription using the integrated system and revealed that it shows unique characteristics, which will be reported elsewhere (Masubuchi et al., in preparation).

4.2 Controlling Biochemical Reaction by Molecular Layout Design

The first step of enzymatic reaction is the collision of enzyme and the substrate, and the collision efficiency depends on the concentration of enzyme and the substrate. Given that the concentration depends on the intermolecular distance, one may assume that the intermolecular distance affects the effective concentration and collision frequency, resulting in the enzymatic activity. Since DNA nanostructure is the versatile method to organize the molecular layout at nanometer resolution, DNA nanostructure is a good platform to examine this hypothesis: controlling the enzymatic activity by intermolecular distance.

Willner and coworkers performed pioneering work on this idea.[150] Using DNA nanostructure of two-hexagon (13 nm width) and four-hexagon (33 nm width), they examined the effect of intermolecular distance on cascade reaction (Fig. 10A). Upon anchoring the two enzymes glucose oxidase (GOx, catalyzes the oxidization of glucose to gluconic acid and produces H_2O_2 concomitantly) and horseradish peroxidase (HRP, consumes H_2O_2 and oxidizes colorless $ABTS^{2-}$ to colored $ABTS^{-}$), the enhancement of cascade reaction compared to freely diffusing enzymes was observed. In addition, the intermolecular distance dependent activity was observed, albeit the difference is small: 13 nm hexagon system shows 120% activity of that of 33 nm hexagon system. This difference is explained by the local concentration difference of H_2O_2: 33 nm hexagon's local concentration is lower than 13 nm hexagon's due to partial diffusion of H_2O_2 to the bulk solution. They further explored the cascade reaction of enzyme (NAD^+-dependent glucose dehydrogenase, GDH) and cofactor NAD^+. Changing the single strand soft linker for tethering cofactor NAD^+ on the DNA nanostructure from 10 to 90 nucleotides, they found that as the soft linker's lengths are shorter, the communication between the enzyme and the cofactor becomes less efficient, and the activity of whole cascade reaction is decreased. Taken together, the cascade reaction is tunable by the intermolecular distance.

Yan and coworkers further extended the Willner's study. They tethered GOx and HRP on DNA origami to evaluate the effect of intermolecular distance ranging from 10 to 65 nm (Fig. 10B),[151] and drew a similar conclusion as Willner's: the closer the distance between the 2 enzymes is, the higher is the activity of HRP. Furthermore, evaluating the cofactor NAD^+'s distance effect into the cascade system of enzyme and cofactor, they found that the molecules anchored 7 nm apart from the enzyme are comparable to the effective concentration of 250 μM, which is steeply decreased to 2.7 μM when placed 14 nm apart (Fig. 10C).[152] Using this steep distance dependence they engineered the pathway with multiple enzymes where the product of the first enzymatic reaction, NADH, was mainly consumed by the tethered second enzyme (malic dehydrogenase, in Fig. 10D), but not by the freely diffusing competitor enzyme (lactate dehydrogenase).

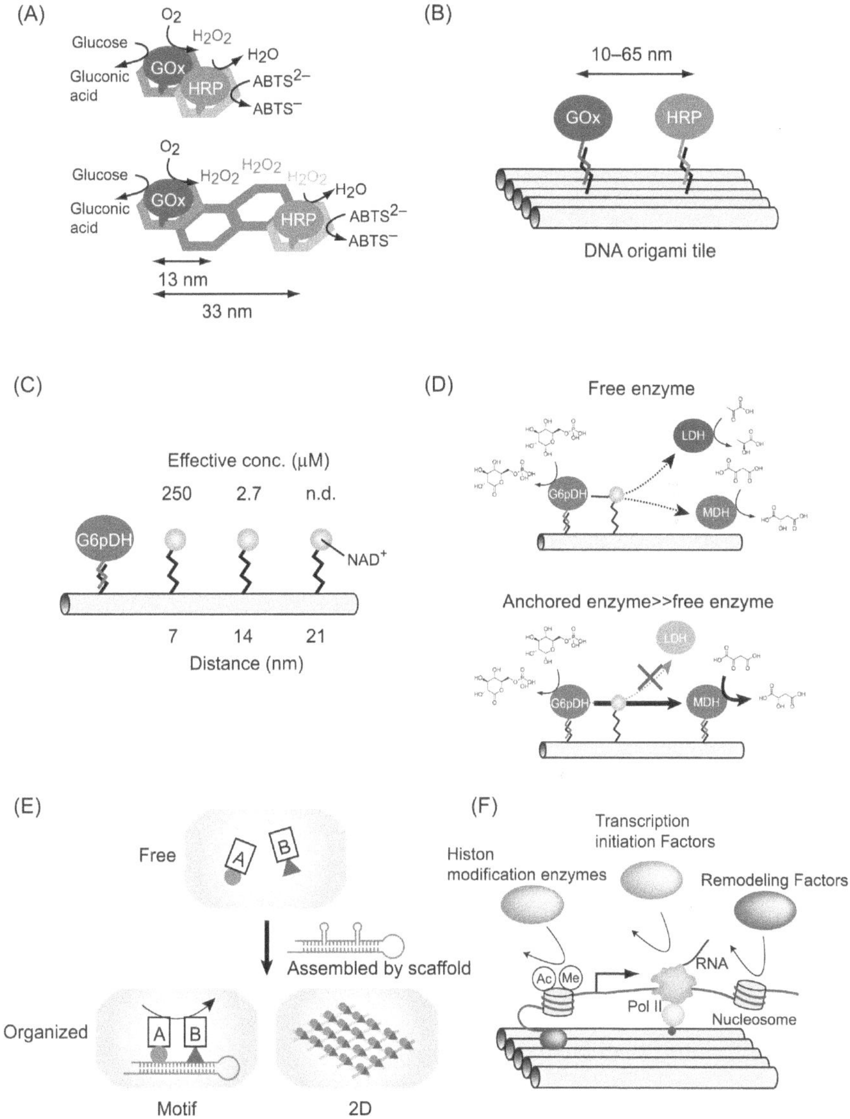

Figure 10 Controlling biochemical reaction by molecular layout design. (A) Cascade enzymatic reaction is enhanced by organizing the intermolecular distance of enzymes, for example, GOx: glucose oxidase and HRP: horseradish peroxidase. (B) Using DNA origami, intermolecular distance effect was examined systematically. (C) Evaluation of the effects of the distance between cofactor NAD^+ and the enzyme (G6pDH). The NAD^+ molecule anchored 7 nm apart from the enzyme is comparable to the effective concentration of 250 μM. When anchored at 14 nm, it is steeply decreased to 2.7 μM. (D) The strong dependence on the distance of effective concentration used to design controlled enzymatic pathway. The product of the first enzymatic reaction, NADH, is

Using an RNA scaffold and RNA-binding proteins such as PP7 and MS2, bacterial metabolism was also rationally designed, where RNA-binding proteins were assembled onto RNA scaffold having aptamer domains and polymerization domains. Thus, the spatial arrangement of cascade enzymes fused to RNA-binding protein could be orchestrated in 1D and 2D. Using this method, the activity of hydrogen-producing pathway was controlled by the architecture of RNA scaffold (Fig. 10E),[153] where PP7-fused [FeFe]-hydrogenase and MS2-fused ferredoxin catalyzes the reduction of protons to hydrogen through electron transfer. Given the in vitro design ability of RNA[154,155] and RNP nanotechnology[156] (see also chapter: RNA and RNP as Building Blocks for Nanotechnology and Synthetic Biology of this book by Saito and coworkers), and the in vivo native architecture of RNP using long noncoding RNA (lncRNA),[157] intermolecular distance control approach would be applicable more in vivo in the near future. To avoid effect on native RNA metabolism, and to ensure the folding process of RNA and RNP nanostructure that are easily disrupted by the existence of RNA-binding protein, the production of RNA and RNP nanostructure via the orthogonal transcription machinery is the key.

Given the intermolecular distance effect on the multiple enzyme system, it might be interesting to address this point on other multiple enzymatic systems such as signal transduction[158] and gene expression system.[159–163]

5. PROSPECTS

Gene expression proceeds in a heterogeneous but highly organized reaction field. Therefore, the layout of the molecules is important. However, there exists little way to control the parameters of molecular layout such as number, order, and intermolecular distance of components. Using DNA nanotechnology, that enables placing molecules in a field nanometer to micrometer range with nanometer resolution at will, would provide an access to address such problems and would be useful to systematically explore

mainly consumed by the tethered second enzyme (MDH: malic dehydrogenase), but not by the freely diffusing competitor enzyme (LDH: lactate dehydrogenase). (E) In vivo enhancement of cascade enzymatic reaction using RNA scaffold and RNA motif binding proteins. (F) Gene nanochip, integrating enzyme (RNAP), and substrate (template gene) on the DNA origami, may allow to evaluate the precise mechanism of the multiple cascade enzymatic gene expression reaction in detail. *(Part A: adapted from Wilner et al. (2009).[150] Part B: adapted from Fu et al. (2012).[151] Part C, D: adapted from Fu et al. (2014).[152] Part E: adapted from Delebecque et al. (2011).)[153]*

the effect of molecular layout. For example, exploring the effect of transcription factors, existing far away from the transcription start site (TSS) or RNAP, is quite challenging but an interesting issue (Fig. 10F).

ACKNOWLEDGMENTS

We thank Hiroshi Sasaki for reading the paper and providing useful comments. This work was supported in part by a Grant-in-Aid for Scientific Research on Innovative Areas ("Functional machinery for non-coding RNAs," "non-coding RNA neo-taxonomy," and "Molecular robotics") (to H.T.), a Grant-in-Aid for Young Scientists (A) (to H.T.), and a Grant-in-Aid for JSPS Fellows to (T.M.).

REFERENCES

1. Tahirov TH, Temiakov D, Anikin M, Patlan V, McAllister WT, Vassylyev DG, Yokoyama S. Structure of a T7 RNA polymerase elongation complex at 2.9 A resolution. *Nature.* 2002;420:43–50.
2. Vassylyev DG, Sekine S, Laptenko O, Lee J, Vassylyeva MN, Borukhov S, Yokoyama S. Crystal structure of a bacterial RNA polymerase holoenzyme at 2.6 A resolution. *Nature.* 2002;417:712–719.
3. Cramer P, Bushnell DA, Kornberg RD. Structural basis of transcription: RNA polymerase II at 2.8 angstrom resolution. *Science.* 2001;292:1863–1876.
4. Gnatt AL, Cramer P, Fu J, Bushnell DA, Kornberg RD. Structural basis of transcription: an RNA polymerase II elongation complex at 3.3 A resolution. *Science.* 2001;292:1876–1882.
5. Plaschka C, Lariviere L, Wenzeck L, Seizl M, Hemann M, Tegunov D, Petrotchenko EV, Borchers CH, Baumeister W, Herzog F, Villa E, Cramer P. Architecture of the RNA polymerase II-Mediator core initiation complex. *Nature.* 2015;518:376–380.
6. Funatsu T, Harada Y, Tokunaga M, Saito K, Yanagida T. Imaging of single fluorescent molecules and individual ATP turnovers by single myosin molecules in aqueous solution. *Nature.* 1995;374:555–559.
7. Ando T, Kodera N, Takai E, Maruyama D, Saito K, Toda A. A high-speed atomic force microscope for studying biological macromolecules. *Proc Natl Acad Sci USA.* 2001;98:12468–12472.
8. Uemura S, Dorywalska M, Lee TH, Kim HD, Puglisi JD, Chu S. Peptide bond formation destabilizes Shine–Dalgarno interaction on the ribosome. *Nature.* 2007;446:454–457.
9. Uemura S, Aitken CE, Korlach J, Flusberg BA, Turner SW, Puglisi JD. Real-time tRNA transit on single translating ribosomes at codon resolution. *Nature.* 2010;464:1012–1017.
10. Tsai A, Petrov A, Marshall RA, Korlach J, Uemura S, Puglisi JD. Heterogeneous pathways and timing of factor departure during translation initiation. *Nature.* 2012;487:390–393.
11. Wang MD, Schnitzer MJ, Yin H, Landick R, Gelles J, Block SM. Force and velocity measured for single molecules of RNA polymerase. *Science.* 1998;282:902–907.
12. Harada Y, Ohara O, Takatsuki A, Itoh H, Shimamoto N, Kinosita Jr K. Direct observation of DNA rotation during transcription by *Escherichia coli* RNA polymerase. *Nature.* 2001;409:113–115.
13. Neuman KC, Abbondanzieri EA, Landick R, Gelles J, Block SM. Ubiquitous transcriptional pausing is independent of RNA polymerase backtracking. *Cell.* 2003;115:437–447.

14. Abbondanzieri EA, Greenleaf WJ, Shaevitz JW, Landick R, Block SM. Direct observation of base-pair stepping by RNA polymerase. *Nature.* 2005;438:460–465.
15. Friedman LJ, Gelles J. Mechanism of transcription initiation at an activator-dependent promoter defined by single-molecule observation. *Cell.* 2012;148:679–689.
16. Fazal FM, Meng CA, Murakami K, Kornberg RD, Block SM. Real-time observation of the initiation of RNA polymerase II transcription. *Nature.* 2015;525:274–277.
17. Kodera N, Yamamoto D, Ishikawa R, Ando T. Video imaging of walking myosin V by high-speed atomic force microscopy. *Nature.* 2010;468:72–76.
18. Uchihashi T, Iino R, Ando T, Noji H. High-speed atomic force microscopy reveals rotary catalysis of rotorless F1-ATPase. *Science.* 2011;333:755–758.
19. Igarashi K, Uchihashi T, Koivula A, Wada M, Kimura S, Okamoto T, Penttila M, Ando T, Samejima M. Traffic jams reduce hydrolytic efficiency of cellulase on cellulose surface. *Science.* 2011;333:1279–1282.
20. Jackson DA, Hassan AB, Errington RJ, Cook PR. Visualization of focal sites of transcription within human nuclei. *EMBO J.* 1993;12:1059–1065.
21. Wansink DG, Schul W, van der Kraan I, van Steensel B, van Driel R, de Jong L. Fluorescent labeling of nascent RNA reveals transcription by RNA polymerase II in domains scattered throughout the nucleus. *J Cell Biol.* 1993;122:283–293.
22. Iborra FJ, Pombo A, Jackson DA, Cook PR. Active RNA polymerases are localized within discrete transcription "factories" in human nuclei. *J Cell Sci.* 1996;109:1427–1436.
23. Jackson DA, Iborra FJ, Manders EM, Cook PR. Numbers and organization of RNA polymerases, nascent transcripts, and transcription units in HeLa nuclei. *Mol Biol Cell.* 1998;9:1523–1536.
24. Pombo A, Jackson DA, Hollinshead M, Wang Z, Roeder RG, Cook PR. Regional specialization in human nuclei: visualization of discrete sites of transcription by RNA polymerase III. *EMBO J.* 1999;18:2241–2253.
25. Osborne CS, Chakalova L, Brown KE, Carter D, Horton A, Debrand E, Goyenechea B, Mitchell JA, Lopes S, Reik W, Fraser P. Active genes dynamically colocalize to shared sites of ongoing transcription. *Nat Genet.* 2004;36:1065–1071.
26. Brown JM, Green J, das Neves RP, Wallace HA, Smith AJ, Hughes J, Gray N, Taylor S, Wood WG, Higgs DR, Iborra FJ, Buckle VJ. Association between active genes occurs at nuclear speckles and is modulated by chromatin environment. *J Cell Biol.* 2008;182:1083–1097.
27. Eskiw CH, Rapp A, Carter DR, Cook PR. RNA polymerase II activity is located on the surface of protein-rich transcription factories. *J Cell Sci.* 2008;121:1999–2007.
28. Eskiw CH, Fraser P. Ultrastructural study of transcription factories in mouse erythroblasts. *J Cell Sci.* 2011;124:3676–3683.
29. Misteli T. Beyond the sequence: cellular organization of genome function. *Cell.* 2007;128:787–800.
30. Hirose T, Mishima Y, Tomari Y. Elements and machinery of non-coding RNAs: toward their taxonomy. *EMBO Rep.* 2014;15:489–507.
31. Maniatis T, Reed R. An extensive network of coupling among gene expression machines. *Nature.* 2002;416:499–506.
32. Seeman NC. Nucleic acid junctions and lattices. *J Theor Biol.* 1982;99:237–247.
33. Fu TJ, Seeman NC. DNA double-crossover molecules. *Biochemistry.* 1993;32:3211–3220.
34. Yan H, Park SH, Finkelstein G, Reif JH, LaBean TH. DNA-templated self-assembly of protein arrays and highly conductive nanowires. *Science.* 2003;301:1882–1884.
35. He Y, Ye T, Su M, Zhang C, Ribbe AE, Jiang W, Mao C. Hierarchical self-assembly of DNA into symmetric supramolecular polyhedra. *Nature.* 2008;452:198–201.
36. Kallenbach NR, Ma RI, Seeman NC. An immobile nucleic acid junction constructed from oligonucleotides. *Nature.* 1983;305:829–831.

37. Winfree E, Liu F, Wenzler LA, Seeman NC. Design and self-assembly of two-dimensional DNA crystals. *Nature.* 1998;394:539–544.
38. Yin P, Choi HM, Calvert CR, Pierce NA. Programming biomolecular self-assembly pathways. *Nature.* 2008;451:318–322.
39. Yin P, Hariadi RF, Sahu S, Choi HM, Park SH, Labean TH, Reif JH. Programming DNA tube circumferences. *Science.* 2008;321:824–826.
40. Wei B, Dai M, Yin P. Complex shapes self-assembled from single-stranded DNA tiles. *Nature.* 2012;485:623–626.
41. Ke Y, Ong LL, Shih WM, Yin P. Three-dimensional structures self-assembled from DNA bricks. *Science.* 2012;338:1177–1183.
42. Rothemund PW. Folding DNA to create nanoscale shapes and patterns. *Nature.* 2006;440:297–302.
43. He Y, Tian Y, Ribbe AE, Mao C. Highly connected two-dimensional crystals of DNA six-point-stars. *J Am Chem Soc.* 2006;128:15978–15979.
44. Chen JH, Seeman NC. Synthesis from DNA of a molecule with the connectivity of a cube. *Nature.* 1991;350:631–633.
45. Zhang Y, Seeman NC. Construction of a DNA-truncated octahedron. *J Am Chem Soc.* 1994;116:1661–1669.
46. Goodman RP, Schaap IA, Tardin CF, Erben CM, Berry RM, Schmidt CF, Turberfield AJ. Rapid chiral assembly of rigid DNA building blocks for molecular nanofabrication. *Science.* 2005;310:1661–1665.
47. Liu H, Chen Y, He Y, Ribbe AE, Mao C. Approaching the limit: can one DNA oligonucleotide assemble into large nanostructures? *Angew Chem Int Ed Engl.* 2006;45:1942–1945.
48. Barish RD, Rothemund PW, Winfree E. Two computational primitives for algorithmic self-assembly: copying and counting. *Nano Lett.* 2005;5:2586–2592.
49. Schulman R, Winfree E. Synthesis of crystals with a programmable kinetic barrier to nucleation. *Proc Natl Acad Sci USA.* 2007;104:15236–15241.
50. Ke Y, Ong LL, Sun W, Song J, Dong M, Shih WM, Yin P. DNA brick crystals with prescribed depths. *Nat Chem.* 2014;6:994–1002.
51. Dunn KE, Dannenberg F, Ouldridge TE, Kwiatkowska M, Turberfield AJ, Bath J. Guiding the folding pathway of DNA origami. *Nature.* 2015;525:82–86.
52. Shih WM, Quispe JD, Joyce GF. A 1.7-kilobase single-stranded DNA that folds into a nanoscale octahedron. *Nature.* 2004;427:618–621.
53. Woo S, Rothemund PW. Programmable molecular recognition based on the geometry of DNA nanostructures. *Nat Chem.* 2011;3:620–627.
54. Sobczak JP, Martin TG, Gerling T, Dietz H. Rapid folding of DNA into nanoscale shapes at constant temperature. *Science.* 2012;338:1458–1461.
55. Castro CE, Kilchherr F, Kim DN, Shiao EL, Wauer T, Wortmann P, Bathe M, Dietz H. A primer to scaffolded DNA origami. *Nat Methods.* 2011;8:221–229.
56. Bellot G, McClintock MA, Lin C, Shih WM. Recovery of intact DNA nanostructures after agarose gel-based separation. *Nat Methods.* 2011;8:192–194.
57. Stahl E, Martin TG, Praetorius F, Dietz H. Facile and scalable preparation of pure and dense DNA origami solutions. *Angew Chem Int Ed Engl.* 2014;53:12735–12740.
58. Torisawa T, Ichikawa M, Furuta A, Saito K, Oiwa K, Kojima H, Toyoshima YY, Furuta K. Autoinhibition and cooperative activation mechanisms of cytoplasmic dynein. *Nat Cell Biol.* 2014;16:1118–1124.
59. Wagenbauer KF, Wachauf CH, Dietz H. Quantifying quality in DNA self-assembly. *Nat Commun.* 2014;5:3691.
60. Douglas SM, Chou JJ, Shih WM. DNA-nanotube-induced alignment of membrane proteins for NMR structure determination. *Proc Natl Acad Sci USA.* 2007;104:6644–6648.

61. Douglas SM, Dietz H, Liedl T, Hogberg B, Graf F, Shih WM. Self-assembly of DNA into nanoscale three-dimensional shapes. *Nature.* 2009;459:414–418.
62. Mathieu F, Liao S, Kopatsch J, Wang T, Mao C, Seeman NC. Six-helix bundles designed from DNA. *Nano Lett.* 2005;5:661–665.
63. Dietz H, Douglas SM, Shih WM. Folding DNA into twisted and curved nanoscale shapes. *Science.* 2009;325:725–730.
64. Andersen ES, Dong M, Nielsen MM, Jahn K, Subramani R, Mamdouh W, Golas MM, Sander B, Stark H, Oliveira CL, Pedersen JS, Birkedal V, Besenbacher F, Gothelf KV, Kjems J. Self-assembly of a nanoscale DNA box with a controllable lid. *Nature.* 2009;459:73–76.
65. Endo M, Hidaka K, Kato T, Namba K, Sugiyama H. DNA prism structures constructed by folding of multiple rectangular arms. *J Am Chem Soc.* 2009;131:15570–15571.
66. Kuzuya A, Komiyama M. Design and construction of a box-shaped 3D-DNA origami. *Chem Commun (Camb).* 2009;28:4182–4184.
67. Han D, Pal S, Nangreave J, Deng Z, Liu Y, Yan H. DNA origami with complex curvatures in three-dimensional space. *Science.* 2011;332:342–346.
68. Han D, Pal S, Yang Y, Jiang S, Nangreave J, Liu Y, Yan H. DNA gridiron nanostructures based on four-arm junctions. *Science.* 2013;339:1412–1415.
69. Benson E, Mohammed A, Gardell J, Masich S, Czeizler E, Orponen P, Hogberg B. DNA rendering of polyhedral meshes at the nanoscale. *Nature.* 2015;523:441–444.
70. Douglas SM, Marblestone AH, Teerapittayanon S, Vazquez A, Church GM, Shih WM. Rapid prototyping of 3D DNA-origami shapes with caDNAno. *Nucleic Acids Res.* 2009;37:5001–5006.
71. Iinuma R, Ke Y, Jungmann R, Schlichthaerle T, Woehrstein JB, Yin P. Polyhedra self-assembled from DNA tripods and characterized with 3D DNA-PAINT. *Science.* 2014;344:65–69.
72. Gerling T, Wagenbauer KF, Neuner AM, Dietz H. Dynamic DNA devices and assemblies formed by shape-complementary, non-base pairing 3D components. *Science.* 2015;347:1446–1452.
73. Reiter NJ, Osterman A, Torres-Larios A, Swinger KK, Pan T, Mondragon A. Structure of a bacterial ribonuclease P holoenzyme in complex with tRNA. *Nature.* 2010;468: 784–789.
74. Endo M, Sugita T, Katsuda Y, Hidaka K, Sugiyama H. Programmed-assembly system using DNA jigsaw pieces. *Chemistry.* 2010;16:5362–5368.
75. Endo M, Sugita T, Rajendran A, Katsuda Y, Emura T, Hidaka K, Sugiyama H. Two-dimensional DNA origami assemblies using a four-way connector. *Chem Commun (Camb).* 2011;47:3213–3215.
76. Wang R, Kuzuya A, Liu W, Seeman NC. Blunt-ended DNA stacking interactions in a 3-helix motif. *Chem Commun (Camb).* 2010;46:4905–4907.
77. Pfitzner E, Wachauf C, Kilchherr F, Pelz B, Shih WM, Rief M, Dietz H. Rigid DNA beams for high-resolution single-molecule mechanics. *Angew Chem Int Ed Engl.* 2013;52:7766–7771.
78. Marchi AN, Saaem I, Vogen BN, Brown S, LaBean TH. Toward larger DNA origami. *Nano Lett.* 2014;14:5740–5747.
79. Zhang H, Chao J, Pan D, Liu H, Huang Q, Fan C. Folding super-sized DNA origami with scaffold strands from long-range PCR. *Chem Commun.* 2012;48:6405–6407.
80. Nickels PC, Ke Y, Jungmann R, Smith DM, Leichsenring M, Shih WM, Liedl T, Hogberg B. DNA origami structures directly assembled from intact bacteriophages. *Small.* 2014;10:1765–1769.
81. Kauert DJ, Kurth T, Liedl T, Seidel R. Direct mechanical measurements reveal the material properties of three-dimensional DNA origami. *Nano Lett.* 2011;11: 5558–5563.

82. Hayashi M, Harada Y. Direct observation of the reversible unwinding of a single DNA molecule caused by the intercalation of ethidium bromide. *Nucleic Acids Res.* 2007;35: e125.
83. Strick TR, Bensimon D, Croquette V. Micro-mechanical measurement of the torsional modulus of DNA. *Genetica.* 1999;106:57–62.
84. Miyazono Y, Hayashi M, Karagiannis P, Harada Y, Tadakuma H. Strain through the neck linker ensures processive runs: a DNA-kinesin hybrid nanomachine study. *EMBO J.* 2010;29:93–106.
85. Derr ND, Goodman BS, Jungmann R, Leschziner AE, Shih WM, Reck-Peterson SL. Tug-of-war in motor protein ensembles revealed with a programmable DNA origami scaffold. *Science.* 2012;338:662–665.
86. Nakata E, Liew FF, Uwatoko C, Kiyonaka S, Mori Y, Katsuda Y, Endo M, Sugiyama H, Morii T. Zinc-finger proteins for site-specific protein positioning on DNA-origami structures. *Angew Chem Int Ed Engl.* 2012;51:2421–2424.
87. Rinker S, Ke Y, Liu Y, Chhabra R, Yan H. Self-assembled DNA nanostructures for distance-dependent multivalent ligand-protein binding. *Nat Nanotechnol.* 2008;3:418–422.
88. Voigt NV, Torring T, Rotaru A, Jacobsen MF, Ravnsbaek JB, Subramani R, Mamdouh W, Kjems J, Mokhir A, Besenbacher F, Gothelf KV. Single-molecule chemical reactions on DNA origami. *Nat Nanotechnol.* 2010;5:200–203.
89. Howarth M, Chinnapen DJ, Gerrow K, Dorrestein PC, Grandy MR, Kelleher NL, El-Husseini A, Ting AY. A monovalent streptavidin with a single femtomolar biotin binding site. *Nat Methods.* 2006;3:267–273.
90. Howarth M, Ting AY. Imaging proteins in live mammalian cells with biotin ligase and monovalent streptavidin. *Nat Protoc.* 2008;3:534–545.
91. Leppiniemi J, Maatta JA, Hammaren H, Soikkeli M, Laitaoja M, Janis J, Kulomaa MS, Hytonen VP. Bifunctional avidin with covalently modifiable ligand binding site. *PLoS One.* 2011;6:e16576.
92. Terai T, Maki E, Sugiyama S, Takahashi Y, Matsumura H, Mori Y, Nagano T. Rational development of caged-biotin protein-labeling agents and some applications in live cells. *Chem Biol.* 2011;18:1261–1272.
93. Herzog VA, Ameres SL. Approaching the golden fleece a molecule at a time: biophysical insights into argonaute-instructed nucleic acid interactions. *Mol Cell.* 2015;59:4–7.
94. Wang Y, Sheng G, Juranek S, Tuschl T, Patel DJ. Structure of the guide-strand-containing argonaute silencing complex. *Nature.* 2008;456:209–213.
95. Wang Y, Juranek S, Li H, Sheng G, Tuschl T, Patel DJ. Structure of an argonaute silencing complex with a seed-containing guide DNA and target RNA duplex. *Nature.* 2008;456:921–926.
96. Wang Y, Juranek S, Li H, Sheng G, Wardle GS, Tuschl T, Patel DJ. Nucleation, propagation and cleavage of target RNAs in Ago silencing complexes. *Nature.* 2009;461:754–761.
97. Wee LM, Flores-Jasso CF, Salomon WE, Zamore PD. Argonaute divides its RNA guide into domains with distinct functions and RNA-binding properties. *Cell.* 2012;151:1055–1067.
98. Yao C, Sasaki HM, Ueda T, Tomari Y, Tadakuma H. Single-molecule analysis of the target cleavage reaction by the *Drosophila* RNAi enzyme complex. *Mol Cell.* 2015;59:125–132.
99. Salomon WE, Jolly SM, Moore MJ, Zamore PD, Serebrov V. Single-molecule imaging reveals that argonaute reshapes the binding properties of its nucleic acid guides. *Cell.* 2015;162:84–95.
100. Deng W, Shi X, Tjian R, Lionnet T, Singer RH. CASFISH: CRISPR/Cas9-mediated in situ labeling of genomic loci in fixed cells. *Proc Natl Acad Sci USA.* 2015;112: 11870–11875.

101. Nihongaki Y, Kawano F, Nakajima T, Sato M. Photoactivatable CRISPR-Cas9 for optogenetic genome editing. *Nat Biotechnol.* 2015;33:755–760.
102. Ran FA, Cong L, Yan WX, Scott DA, Gootenberg JS, Kriz AJ, Zetsche B, Shalem O, Wu X, Makarova KS, Koonin EV, Sharp PA, Zhang F. In vivo genome editing using *Staphylococcus aureus* Cas9. *Nature.* 2015;520:186–191.
103. Nishimasu H, Cong L, Yan WX, Ran FA, Zetsche B, Li Y, Kurabayashi A, Ishitani R, Zhang F, Nureki O. Crystal structure of *Staphylococcus aureus* Cas9. *Cell.* 2015;162: 1113–1126.
104. Zetsche B, Gootenberg JS, Abudayyeh OO, Slaymaker IM, Makarova KS, Essletzbichler P, Volz SE, Joung J, van der Oost J, Regev A, Koonin EV, Zhang F. Cpf1 is a single RNA-guided endonuclease of a Class 2 CRISPR-Cas system. *Cell.* 2015;163:759–771.
105. Konermann S, Brigham MD, Trevino AE, Joung J, Abudayyeh OO, Barcena C, Hsu PD, Habib N, Gootenberg JS, Nishimasu H, Nureki O, Zhang F. Genome-scale transcriptional activation by an engineered CRISPR-Cas9 complex. *Nature.* 2015;517(7536):583–588.
106. Shechner DM, Hacisuleyman E, Younger ST, Rinn JL. Multiplexable, locus-specific targeting of long RNAs with CRISPR-display. *Nat Methods.* 2015;12:664–670.
107. Chhabra R, Sharma J, Ke Y, Liu Y, Rinker S, Lindsay S, Yan H. Spatially addressable multiprotein nanoarrays templated by aptamer-tagged DNA nanoarchitectures. *J Am Chem Soc.* 2007;129:10304–10305.
108. Kimoto M, Yamashige R, Matsunaga K, Yokoyama S, Hirao I. Generation of high-affinity DNA aptamers using an expanded genetic alphabet. *Nat Biotechnol.* 2013;31: 453–457.
109. He Y, Tian Y, Ribbe AE, Mao C. Antibody nanoarrays with a pitch of approximately 20 nanometers. *J Am Chem Soc.* 2006;128:12664–12665.
110. Williams BA, Lund K, Liu Y, Yan H, Chaput JC. Self-assembled peptide nanoarrays: an approach to studying protein–protein interactions. *Angew Chem Int Ed Engl.* 2007;46:3051–3054.
111. Kuzuya A, Sakai Y, Yamazaki T, Xu Y, Komiyama M. Nanomechanical DNA origami 'single-molecule beacons' Directly Imaged By Atomic Force microscopy. *Nat Commun.* 2011;2:449.
112. Osada E, Suzuki Y, Hidaka K, Ohno H, Sugiyama H, Endo M, Saito H. Engineering RNA–protein complexes with different shapes for imaging and therapeutic applications. *ACS Nano.* 2014;8:8130–8140.
113. Arbabi Ghahroudi M, Desmyter A, Wyns L, Hamers R, Muyldermans S. Selection and identification of single domain antibody fragments from camel heavy-chain antibodies. *FEBS Lett.* 1997;414:521–526.
114. Nord K, Nilsson J, Nilsson B, Uhlen M, Nygren PA. A combinatorial library of an alpha-helical bacterial receptor domain. *Protein Eng.* 1995;8:601–608.
115. Nord K, Gunneriusson E, Ringdahl J, Stahl S, Uhlen M, Nygren PA. Binding proteins selected from combinatorial libraries of an alpha-helical bacterial receptor domain. *Nat Biotechnol.* 1997;15:772–777.
116. Hamers-Casterman C, Atarhouch T, Muyldermans S, Robinson G, Hamers C, Songa EB, Bendahman N, Hamers R. Naturally occurring antibodies devoid of light chains. *Nature.* 1993;363:446–448.
117. Harmsen MM, De Haard HJ. Properties, production, and applications of camelid single-domain antibody fragments. *Appl Microbiol Biotechnol.* 2007;77:13–22.
118. van der Linden RH, Frenken LG, de Geus B, Harmsen MM, Ruuls RC, Stok W, de Ron L, Wilson S, Davis P, Verrips CT. Comparison of physical chemical properties of llama VHH antibody fragments and mouse monoclonal antibodies. *Biochim Biophys Acta.* 1999;1431:37–46.

119. Orlova A, Magnusson M, Eriksson TL, Nilsson M, Larsson B, Hoiden-Guthenberg I, Widstrom C, Carlsson J, Tolmachev V, Stahl S, Nilsson FY. Tumor imaging using a picomolar affinity HER2 binding affibody molecule. *Cancer Res.* 2006;66:4339–4348.
120. Shen W, Zhong H, Neff D, Norton ML. NTA directed protein nanopatterning on DNA origami nanoconstructs. *J Am Chem Soc.* 2009;131:6660–6661.
121. Goodman RP, Erben CM, Malo J, Ho WM, McKee ML, Kapanidis AN, Turberfield AJ. A facile method for reversibly linking a recombinant protein to DNA. *ChemBioChem.* 2009;10:1551–1557.
122. Lata S, Gavutis M, Tampe R, Piehler J. Specific and stable fluorescence labeling of histidine-tagged proteins for dissecting multi-protein complex formation. *J Am Chem Soc.* 2006;128:2365–2372.
123. Cohen JD, Sadowski JP, Dervan PB. Addressing single molecules on DNA nanostructures. *Angew Chem Int Ed Engl.* 2007;46:7956–7959.
124. Cohen JD, Sadowski JP, Dervan PB. Programming multiple protein patterns on a single DNA nanostructure. *J Am Chem Soc.* 2008;130:402–403.
125. Dervan PB, Edelson BS. Recognition of the DNA minor groove by pyrrole-imidazole polyamides. *Curr Opin Struct Biol.* 2003;13:284–299.
126. Los GV, Wood K. The HaloTag: a novel technology for cell imaging and protein analysis. *Methods Mol Biol.* 2007;356:195–208.
127. Keppler A, Gendreizig S, Gronemeyer T, Pick H, Vogel H, Johnsson K. A general method for the covalent labeling of fusion proteins with small molecules in vivo. *Nat Biotechnol.* 2003;21:86–89.
128. Sacca B, Meyer R, Erkelenz M, Kiko K, Arndt A, Schroeder H, Rabe KS, Niemeyer CM. Orthogonal protein decoration of DNA origami. *Angew Chem Int Ed Engl.* 2010;49 (49):9378–9383.
129. Tornoe CW, Christensen C, Meldal M. Peptidotriazoles on solid phase: [1,2,3]-triazoles by regiospecific copper(i)-catalyzed 1,3-dipolar cycloadditions of terminal alkynes to azides. *J Org Chem.* 2002;67:3057–3064.
130. Rostovtsev VV, Green LG, Fokin VV, Sharpless KB. A stepwise huisgen cycloaddition process: copper(I)-catalyzed regioselective "ligation" of azides and terminal alkynes. *Angew Chem Int Ed Engl.* 2002;41:2596–2599.
131. Chan TR, Hilgraf R, Sharpless KB, Fokin VV. Polytriazoles as copper(I)-stabilizing ligands in catalysis. *Org Lett.* 2004;6:2853–2855.
132. Voigt NV, Torring T, Rotaru A, Jacobsen MF, Ravnsbaek JB, Subramani R, Mamdouh W, Kjems J, Mokhir A, Besenbacher F, Gothelf KV. Single-molecule chemical reactions on DNA origami. *Nat Nanotechnol.* 2010;5:200–203.
133. Baskin JM, Prescher JA, Laughlin ST, Agard NJ, Chang PV, Miller IA, Lo A, Codelli JA, Bertozzi CR. Copper-free click chemistry for dynamic in vivo imaging. *Proc Natl Acad Sci USA.* 2007;104:16793–16797.
134. Devaraj NK, Weissleder R, Hilderbrand SA. Tetrazine-based cycloadditions: application to pretargeted live cell imaging. *Bioconjug Chem.* 2008;19:2297–2299.
135. Devaraj NK, Upadhyay R, Haun JB, Hilderbrand SA, Weissleder R. Fast and sensitive pretargeted labeling of cancer cells through a tetrazine/trans-cyclooctene cycloaddition. *Angew Chem Int Ed Engl.* 2009;48:7013–7016.
136. Blackman ML, Royzen M, Fox JM. Tetrazine ligation: fast bioconjugation based on inverse-electron-demand Diels–Alder reactivity. *J Am Chem Soc.* 2008;130: 13518–13519.
137. Iwasaki S, Sasaki HM, Sakaguchi Y, Suzuki T, Tadakuma H, Tomari Y. Defining fundamental steps in the assembly of the *Drosophila* RNAi enzyme complex. *Nature.* 2015;521:533–536.
138. Betzig E, Chichester RJ. Single molecules observed by near-field scanning optical microscopy. *Science.* 1993;262:1422–1425.

139. Levene MJ, Korlach J, Turner SW, Foquet M, Craighead HG, Webb WW. Zero-mode waveguides for single-molecule analysis at high concentrations. *Science.* 2003;299: 682–686.
140. Acuna GP, Moller FM, Holzmeister P, Beater S, Lalkens B, Tinnefeld P. Fluorescence enhancement at docking sites of DNA-directed self-assembled nanoantennas. *Science.* 2012;338:506–510.
141. Suzuki Y, Endo M, Katsuda Y, Ou K, Hidaka K, Sugiyama H. DNA origami based visualization system for studying site-specific recombination events. *J Am Chem Soc.* 2014;136:211–218.
142. Endo M, Tatsumi K, Terushima K, Katsuda Y, Hidaka K, Harada Y, Sugiyama H. Direct visualization of the movement of a single T7 RNA polymerase and transcription on a DNA nanostructure. *Angew Chem Int Ed Engl.* 2012;51:8778–8782.
143. Rajendran A, Endo M, Sugiyama H. Structural functional analysis of proteins by high-speed atomic force microscopy. *Adv Protein Chem Struct Biol.* 2012;87:5–55.
144. Pibiri E, Holzmeister P, Lalkens B, Acuna GP, Tinnefeld P. Single-molecule positioning in zeromode waveguides by DNA origami nanoadapters. *Nano Lett.* 2014;14:3499–3503.
145. Holzmeister P, Pibiri E, Schmied JJ, Sen T, Acuna GP, Tinnefeld P. Quantum yield and excitation rate of single molecules close to metallic nanostructures. *Nat Commun.* 2014;5:5356.
146. Radmacher M, Fritz M, Hansma HG, Hansma PK. Direct observation of enzyme activity with the atomic force microscope. *Science.* 1994;265:1577–1579.
147. Bustamante C, Marko JF, Siggia ED, Smith S. Entropic elasticity of lambda-phage DNA. *Science.* 1994;265:1599–1600.
148. Rajendran A, Endo M, Sugiyama H. State-of-the-art high-speed atomic force microscopy for investigation of single-molecular dynamics of proteins. *Chem Rev.* 2014;114: 1493–1520.
149. Endo M, Katsuda Y, Hidaka K, Sugiyama H. Regulation of DNA methylation using different tensions of double strands constructed in a defined DNA nanostructure. *J Am Chem Soc.* 2010;132:1592–1597.
150. Wilner OI, Weizmann Y, Gill R, Lioubashevski O, Freeman R, Willner I. Enzyme cascades activated on topologically programmed DNA scaffolds. *Nat Nanotechnol.* 2009;4:249–254.
151. Fu J, Liu M, Liu Y, Woodbury NW, Yan H. Interenzyme substrate diffusion for an enzyme cascade organized on spatially addressable DNA nanostructures. *J Am Chem Soc.* 2012;134:5516–5519.
152. Fu J, Yang YR, Johnson-Buck A, Liu M, Liu Y, Walter NG, Woodbury NW, Yan H. Multi-enzyme complexes on DNA scaffolds capable of substrate channelling with an artificial swinging arm. *Nat Nanotechnol.* 2014;9:531–536.
153. Delebecque CJ, Lindner AB, Silver PA, Aldaye FA. Organization of intracellular reactions with rationally designed RNA assemblies. *Science.* 2011;333:470–474.
154. Guo P. The emerging field of RNA nanotechnology. *Nat Nanotechnol.* 2010;5:833–842.
155. Geary C, Rothemund PW, Andersen ES. RNA nanostructures. A single-stranded architecture for cotranscriptional folding of RNA nanostructures. *Science.* 2014;345: 799–804.
156. Ohno H, Kobayashi T, Kabata R, Endo K, Iwasa T, Yoshimura SH, Takeyasu K, Inoue T, Saito H. Synthetic RNA–protein complex shaped like an equilateral triangle. *Nat Nanotechnol.* 2011;6:116–120.
157. Sasaki YT, Ideue T, Sano M, Mituyama T, Hirose T. MENepsilon/beta noncoding RNAs are essential for structural integrity of nuclear paraspeckles. *Proc Natl Acad Sci USA.* 2009;106:2525–2530.
158. Park SH, Zarrinpar A, Lim WA. Rewiring MAP kinase pathways using alternative scaffold assembly mechanisms. *Science.* 2003;299:1061–1064.

159. Cook PR. The organization of replication and transcription. *Science.* 1999;284: 1790–1795.
160. Sutherland H, Bickmore WA. Transcription factories: gene expression in unions? *Nat Rev Genet.* 2009;10:457–466.
161. Allen BL, Taatjes DJ. The mediator complex: a central integrator of transcription. *Nat Rev Mol Cell Biol.* 2015;16:155–166.
162. Amit R, Garcia HG, Phillips R, Fraser SE. Building enhancers from the ground up: a synthetic biology approach. *Cell.* 2011;146:105–118.
163. Redden H, Alper HS. The development and characterization of synthetic minimal yeast promoters. *Nat Commun.* 2015;6:7810.

CHAPTER SIX

RNA and RNP as Building Blocks for Nanotechnology and Synthetic Biology

Hirohisa Ohno, Hirohide Saito*
Center for iPS Cell Research and Application, Kyoto University, Kyoto, Japan
* Corresponding author. E-mail address: hirohide.saito@cira.kyoto-u.ac.jp

Contents

Abstract

Recent technologies that aimed to elucidate cellular function have revealed essential roles for RNA molecules in living systems. Our knowledge concerning functional and structural information of naturally occurring RNA and RNA-protein (RNP) complexes is increasing rapidly. RNA and RNP interaction motifs are structural units that function as building blocks to constitute variety of complex structures. RNA-central synthetic biology and nanotechnology are constructive approaches that employ the accumulated information and build synthetic RNA (RNP)-based circuits and nanostructures.

Progress in Molecular Biology and Translational Science, Volume 139
ISSN 1877-1173
http://dx.doi.org/10.1016/bs.pmbts.2015.12.004

Here, we describe how to design and construct synthetic RNA (RNP)-based devices and structures at the nanometer-scale for biological and future therapeutic applications. RNA/RNP nanostructures can also be utilized as the molecular scaffold to control the localization or interactions of target molecule(s). Moreover, RNA motifs recognized by RNA-binding proteins can be applied to make protein-responsive translational "switches" that can turn gene expression "on" or "off" depending on the intracellular environment. This "synthetic RNA and RNP world" will expand tools for nanotechnology and synthetic biology. In addition, these reconstructive approaches would lead to a greater understanding of building principle in naturally occurring RNA/RNP molecules and systems.

1. RNA NANOTECHNOLOGY

1.1 RNA as a Nanomaterial

Biomolecules are often used as materials in the nanotechnology field. In particular, DNA has been widely investigated in expanding the field because of its molecular recognition and self-organization ability based on complementary Watson–Crick base pairing. DNA nanotechnology pioneered by Seeman[1,2] has been accelerated by the following report of DNA origami by Rothemund.[3] DNA origami technology enables us to design and construct a variety of two-dimensional (2D) and three-dimensional (3D) nano-scale structures (nanostructures) made of only DNA with high accuracy and efficiency.[4] Recently, in addition to DNA, bioengineers have also been attracted to the use of RNA.[5,6] In a similar manner to DNA molecules, RNA is also able to form Watson–Crick base pairing, and regulate gene expression through DNA–RNA and RNA–RNA interactions. However, the role of RNA is not limited to simple nucleic acid interaction. For example, transfer RNA (tRNA) molecules attached to a corresponding amino acid at their3'-terminus (reaction catalyzed by aminoacyl-tRNA synthetases), and ribosomal RNA plays central roles to catalyze peptide-bond formation within ribosome complex. Furthermore, discovery of various RNA molecules such as ribozymes and riboswitches supports important and versatile functions of RNA in living systems. Why can RNA molecules play variety of functions? Structural analysis of RNA revealed that functional diversity of RNAs is based on their complex 3D structures like proteins. Importantly, these complex RNA structures are made by the self-assembly of many smaller and unorthodox interactions between base–base, base–sugar and base–phosphate. Such distinctive interactive regions with structures are

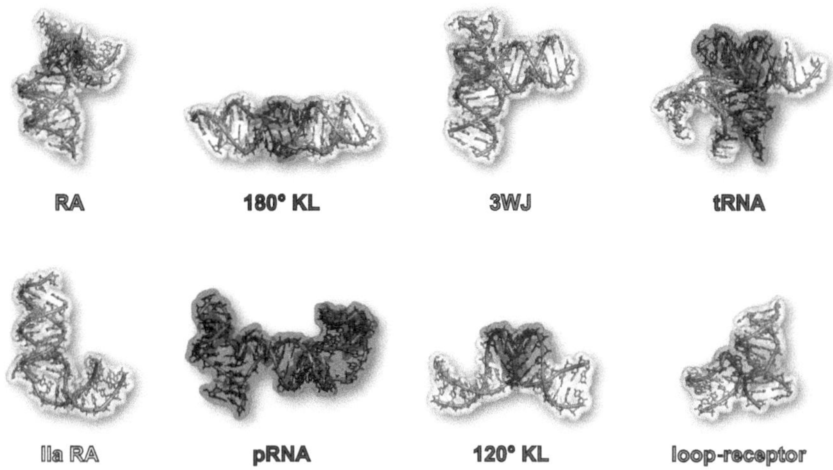

Figure 1 RNA structural motifs. Various RNA structural motifs have been found out from natural RNA molecules. Each 3D data was obtained from PDB (id is shown later). right-angle *(RA)* motif (extracted from 1JJ2[8]); 180° kissing-loop *(KL)* motif (1JJM[9]); three-way junction *(3WJ)* motif (extracted from 4V4Q[10]); *tRNA* motif (extracted from 4TNA[11]); *IIa RA* motif (2NOK[12]); *pRNA* motif (3R4F[13]); 120° *KL* motif (2BJ2[14]); *loop-receptor* motif (extracted from 1HR2[15]). Core region of each motif is colored and white region corresponds to normal Watson–Crick type double-stranded helix region. All of them are drawn in the same scale.

called RNA structural motifs (RNA motifs). So far, a variety of RNA motifs have been found in nature and extracted as building blocks (Fig. 1) to construct artificial nanostructures.[7]

1.2 RNA Nanostructures

Up to now, variety of RNA nanostructures have been constructed and reported, as shown in Fig. 2A and B. One of the earliest and important achievement is the square-shaped nanostructure reported by Chworos et al. (Fig. 2A, I and II).[16] They used an 11-nt right-angle motif from ribosomal RNA (RA motif; Fig. 1) to form the corner of the square and a kissing-loop motif from human immunodeficiency virus type 1 kissing complex (180°KL motif; Fig. 1) to connect the four corners. Each square contains the 3'-protruding tail on each corner. Via the tails, squares can be connected with each other and assemble into planer structures like a ladder or lattice pattern for which they called "RNA jigsaw puzzle."

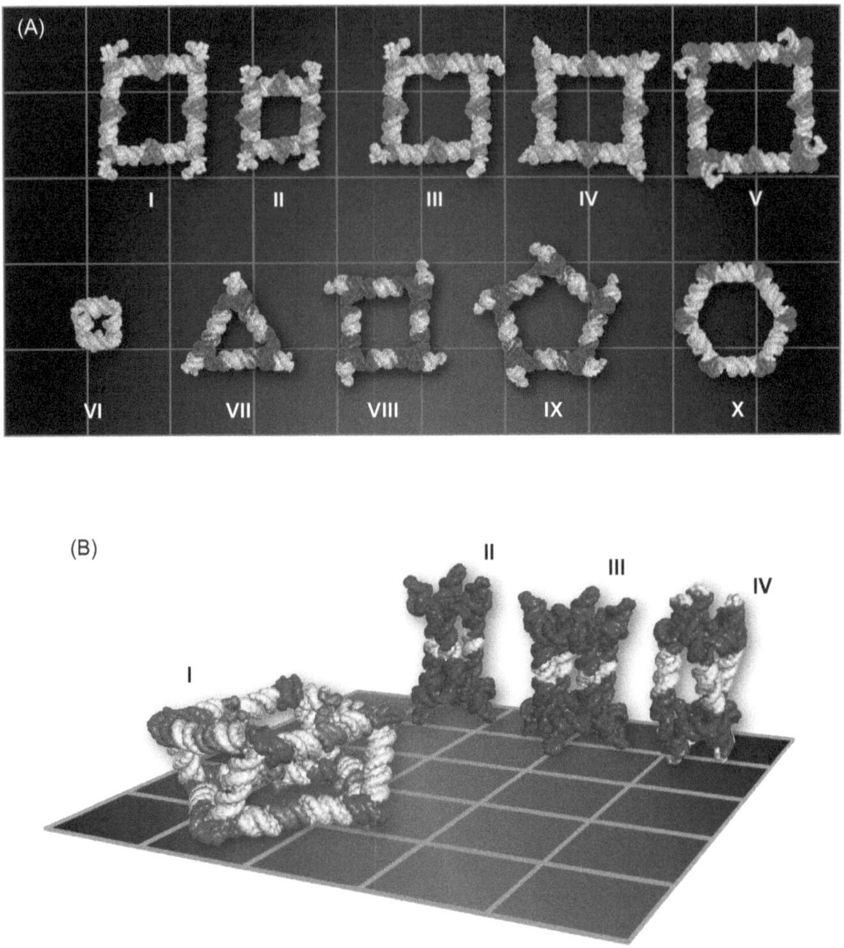

Figure 2 RNA nanostructures. (A) Various 2D and (B) 3D nanostructures have been built with RNA. The 3D models of IIa RA square is obtained from PDB (id: 3P59[17]). Other models are built according to each reference. The mesh size of grid is 10 × 10 nm.

Other types of square-shaped nanostructures have been designed using different RNA motifs. Severcan et al. have built three kinds of square-like structures (Fig. 2A, III–V) using two alternative motifs (3WJ motif and tRNA motif) and RA motif.[18] 3WJ and tRNA motifs (Fig. 1) were extracted from the 50S ribosome of *E. coli* and tRNA, respectively. The formation of these square-shape nanostructures was confirmed by atomic force microscopy (AFM) imaging.

Dibrov et al. extracted a right-angle motif, domain IIa bulge (IIa RA motif; Fig. 1) from the internal ribosomal entry site of hepatitis C virus.[17] The four bent parts consisting of the motif were linked with complementary tails forming a square-shaped structure (Fig. 2A, VI). They determined the square's 3D structure at atomic level by X-ray crystallography.

Guo and coworkers have made several structures using packaging RNA motif (pRNA motif; Fig. 1) from phi29 bacteriophage. pRNA has three loops and two three-way junction regions and forms a hexamer through the loop–loop interaction to compose the DNA packaging machinery of the bacteriophage. By using the loop–loop interaction they created a trimeric complex with a triangular shape.[19] Also, by using the 3-way junction at the corner, three kinds of polygons; triangle, square, and pentagon were constructed (Fig. 2A, VII-IX).[20]

Grabow et al. built a hexagonal nanoring (Fig. 2A, X) using RNAI/II inverse kissing-loop motif with a bent angle at about 120 degrees (120° KL motif; Fig. 1).[21] Monomeric RNA forms a dumbbell-like structure that has loop regions at both ends. This RNA corresponds to each side of the hexagon. Six RNAs are connected by kissing-loop interactions at 120 degrees, resulting in a hexagonal structure. When the motif was replaced by a kissing-loop motif with 180 degrees, RNAs were connected linearly instead of forming the ring.

In addition to the planer polygons, 3D structures have also been designed and constructed. Severcan et al. has built a quadrangular prism (Fig. 2B, I) using a tRNAmotif.[22] Its shape was like a twisted and squashed cube. The square composing top and bottom faces was made with four tRNA motifs and four 180 degrees KL motifs just like the square previously mentioned. The two squares were associated by 180 degrees KL motif or tail–tail interaction, and formed the twisted cube. It is difficult to observe such 3D structures clearly by AFM. To confirm its 3D structure they used cryogenic electron microscopy with single particle analysis instead of AFM, which is more suited to study planar structures.

Hao et al. made three kinds of prisms (Fig. 2B, II-IV) with pRNA.[23] They altered the sequence in the loop region of pRNA motif and controlled the loop–loop interacting pattern. Via the loop–loop interacting regions, two triangular prisms and one tetragonal prism were assembled from six or eight RNAs, respectively. The constructed prisms were confirmed by electrophoretic mobility shift assay (EMSA) and Cryo-EM.

Most of RNA nanostructures are smaller (from several nanometers to several tens of nanometers) than DNA-made nanostructures (generally from several tens of nanometers to hundreds of nanometers), due to the small sizes of RNA motifs to build. However, larger nanostructures can also be built with RNA by self-assembly or origami technology. The RNA-jigsaw puzzle previously discussed is an example of larger RNA nanostructures made by self-assembly (reaches to micrometer-scale).

A filamentous structure was also built by Nasalean et al.[24] They designed an H-shaped molecule based on a four-way junction motif. Loop-receptor motif (Fig. 1) was used to assemble them.

Recently, Geary et al. knit a sheet of tiles from transcribed single-stranded RNA by using kissing-loop and dovetail seam motifs.[25] It has multiple additional kissing-loop motifs on the sides of the tile to connect each tile, thereby creating a 2D honeycomb-like network structure. They observed self-folding of RNA to the tile during the transcription, without the requirement of denaturing-to-renaturing processes.

It is obvious that the same strategy to design DNA nanostructures can be applied to form RNA nanostructures. Indeed, RNA cube was made by simple Watson–Crick base pairing without RNA structural motifs.[26] Six sequence-designed RNA strands were folded into a cube by proper hybridization. Furthermore a variation built from10 strands was also able to form a cube. Termini of the strand can be elongated and used as toe-hold to add any sequence.

Yu et al. also applied a methodology used in DNA nanotechnology to build RNA nanostructures.[27] They combined the eight homologous units by using a T-shaped junction,[28] and built a homo-octameric cube-like prism.

As described earlier, many nanostructures have been designed and built by building principles unique to RNA (RNA motifs) or similar to DNA (Watson–Crick base-paring). These nanostructures can be used as a scaffold to attach desired functions for biological applications.

1.3 Application of RNA Nanostructures

RNA nanostructures may become a carrier for functional molecules because of their compatibility with cellular systems. For example, Functional RNAm molecules such as RNA aptamers (specific binders to target molecules) or short interfering RNA (siRNA), can be loaded onto RNA nanostructures directly and easily. For example, Afonin et al. succeeded in attaching an antidye aptamer and its fluorescent ability to the RNA cube.[26] They also introduced the siRNA-carrying hexagonal nanoring into human cultured cells and into mice.[29] In a similar manner, Grabow et al. introduced siRNA into the hexagonal nanoring.[21] In both cases, the expression of target genes was efficiently inhibited. These RNA nanostructures improved stability of RNA against ribonuclease degradation, suggesting their potential use for drug delivery in vivo. They also installed an aptamer (J18) for antihuman epidermal growth factor (EGFR) into hexagonal nanoring.[29] This chimeric nanoring showed effective binding to target cells that express EGFR at high-levels. Similarly, Guo et al. installed ribozyme, siRNA, or aptamers into the pRNA-based nanostructures.[30,31]

DNA strands can be simply introduced into RNA nanostructures by DNA–RNA hybridization between complementary strands. Khisamutdinov et al. incorporated CpG oligodeoxynucleotide (CpG ODN) to the vertexes of pRNA-polygons.[20] CpG ODN is recognized by toll-like receptor 9 (TLR9) on the macrophage and ingested. CpG-carrying nanostructures showed that the cellular uptake and immunostimulation were dependent on the number of CpG ODNs.

Small molecules can be loaded on RNA nanostructures by chemical conjugation techniques. For instance, fluorescent dyes are popularly used to label nucleic acids. In the researches described earlier, various dyes (FITC[19], Cy3[20,30], Alexa546 and IRDye 700[29]) have been used. Another example is with folate, that is known to be a specific delivery agent to cancer cells. Guo and his coworkers incorporated folate into the RNA nanostructures made with pRNA by hybridization with folate-conjugated DNA.[30,32] They showed that the folate-carrying nanostructures facilitated the cancer cell specific targeting in vitro (cultured cell) and in vivo (mouse).

RNA nanostructure can also carry tagged protein by using specific molecular conjugation. One of the most used tags is streptavidin that binds strongly to biotin. Afonin et al. used a biotinylated DNA strand, and loaded a streptavidin fused red fluorescent protein, phycoerythrin onto the hexagonal ring.[29]

Thus, multifunctional complexes can be assembled easily through the introduction of functional molecules onto the RNA nanostructures. Future applications in drug delivery and nanomedicine are expected.

2. RNA-PROTEIN (RNP) NANOSTRUCTURES

2.1 RNA-Protein Interacting Motifs (RNP Motifs)

Naturally occurring RNAs often form complexes with RNA-binding proteins (RBPs). Such RNP complexes have complicated structures and sophisticated functions. Ribosome is a typical example of highly complex RNPs.[10,33–35] These complex RNP structures can be formed from the assembly of a set of RNP interacting motifs (RNP motifs). RNP motifs may be extracted and used as a building block to reconfigure molecules. Recently, we succeeded to design and construct two different RNP nanostructures, such as triangular-or square-shaped, by using RNP motifs as a building bock.

2.2 L7-box C/D Triangle

Ohno et al. constructed equilateral triangular nanostructures using box C/D kink-turn RNA motif and its binding L7Ae protein (Fig. 3A and B).[36] L7Ae selectively binds box C/D RNA, and bends the kink-turn region at approximately 60 degrees (Fig. 3A), although the RNA is flexible in the absence of L7Ae.[37] This structural change is referred as "induced fit." We designed an equilateral triangle with the module as a building block suitable for the corner and a size length of approximately 17 nm (Fig. 3B).

After the preparation of designed RNA and L7Ae protein, the RNP interaction was confirmed by EMSA. The result showed that the three L7Ae proteins were bound to one unit of RNA, indicating that the designed RNP complex was formed. The configuration of the complex was directly observed using AFM. In the absence of L7Ae, RNA showed heterogeneous structures. In contrast, triangular-like shapes were observed successfully in the presence of L7Ae protein. Later, the configuration was visualized more clearly using high-speed AFM (HS-AFM).[38] The higher resolution images revealed each of RNA-bound proteins as a distinguishable particle. Moreover, the binding process of L7Ae proteins into RNA nanostructures in solution was observed in real-time.

2.3 L1-rRNA Square

Ohno et al. recently developed a square-shaped RNP nanostructure.[39] We used an RNP motif consisting of a ribosomal protein L1 and its binding motif in ribosomal RNA (rRNA motif) (Fig. 3C). It is known that L1 can bind to the rRNA motif and another motif in its mRNA (mRNA motif).[40] In both motifs, L1 induces RNA structural change, generating the bent shape like a right angle. We chose the rRNA motif due to higher affinity to L1 compared with that of the mRNA motif,[41] and designed the square-shaped nanostructure using the rRNA-L1 motif. Two types of nanostructures were designed: one was the unilateral type in which all four proteins

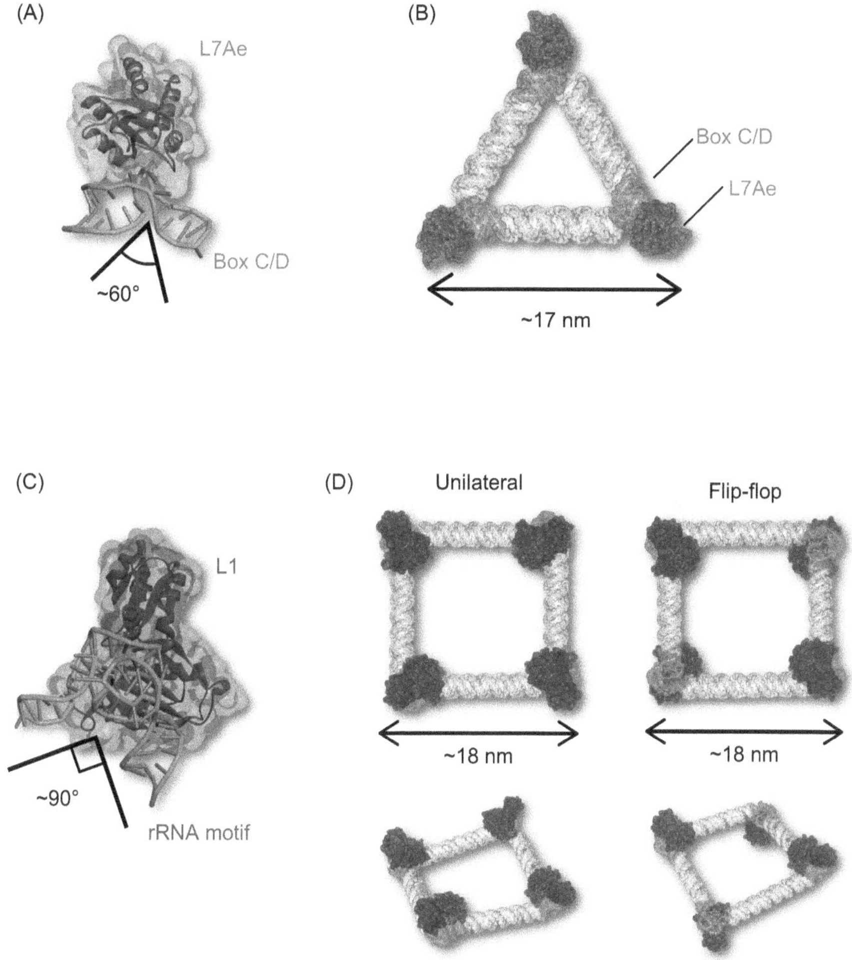

Figure 3 RNP nanostructures. (A) Box C/D-L7Ae module. Box C/D RNA is bent and fixed at ~60 degrees by the bound L7Ae protein; (B) 3D model of L7-triangle. Colored region at apex corresponds to RNP module. White regions consist of ordinary double-stranded RNA; (C) rRNA motif-L1 module. The rRNA motif forms the right angle shape in the complex with L1 protein; and (D) 3D models of two-types of squares.

placed on the one side and another was the flip-flop type in which each face had two proteins respectively (Fig. 3D). They had almost the same size (~18 nm at the side).

The complex formation between the RNA and L1 protein was checked by EMSA and square-shaped particles were observed in the presence of L1 protein by using AFM, although atypical shapes were observed in RNA alone. HS-AFM observation also showed the square-shaped complexes containing four L1 proteins in the unilateral type. However, no complete complex containing four L1 proteins was observed in flip-flop type. This difference in the appearance seems to reflect their 3D structures, in other words, the flatness. Therefore, these results implied that the square-shaped RNP nanostructures were constructed accurately as designed.

2.4 Improvement of RNA Stability in RNP Nanostructure

In general, RNA is thought to be a labile molecule because of its sensitivity to nucleases and degradation in alkaline conditions. Improvement of RNA stability is an important matter for biological application, especially to apply them for living system. To improve RNA stability, chemically modified nucleotides (eg, phosphorothioate nucleotide[42] and 2'-deoxyfluoro ribose nucleotide[43]) are often used to inhibit degradation by nucleases.[44] However, insertion of modified nucleotides into RNA may disrupt molecular interactions, and influence the innate structure and function. In fact, introduction of 2'-deoxyfluoro ribose into box C/D led to the loss of the binding ability to L7Ae.[45] Thus, the usability of modified nucleotides is limited in some cases.

We found that specific RNA–protein interaction (ie, L7Ae-boxC/D and L1-rRNA) can improve stability of the RNA nanostructures without using chemical modifications. To examine the stability, RNP nanostructures were incubated in 20% fetal bovine serum-containing buffer at 37°C. On the triangular structure, RNA was completely degraded within 30 min without L7Ae protein. On the other hand, more than 50% of RNA was retained after 120 min in the presence of L7Ae.[38] The protective effect against nucleases was also observed in L1-square. RNA without L1 protein was degraded completely after 8 h of incubation. In contrast, more than 50% of RNA with L1 remained after 24 h.[39] Such improved stability by the complex formation with protein is one of the advantages to use RNP for biological applications.

3. APPLICATIONS OF RNP NANOSTRUCTURE

3.1 As a Carrier for Functional RNA

In principle, RNP nanostructure could be used as a carrier for any functional molecule. Functional RNA molecules can be attached directly onto the RNA region in a similar manner to RNA nanostructures. Osada et al.

applied L7-triangle as a siRNA carrier.[38] siRNA was loaded onto the sides of the triangle (Fig. 4A, left). First, it was checked whether the siRNA was subjected to processing by Dicer. The result of in vitro assay showed that the corresponding siRNA was produced from this RNP complex. The

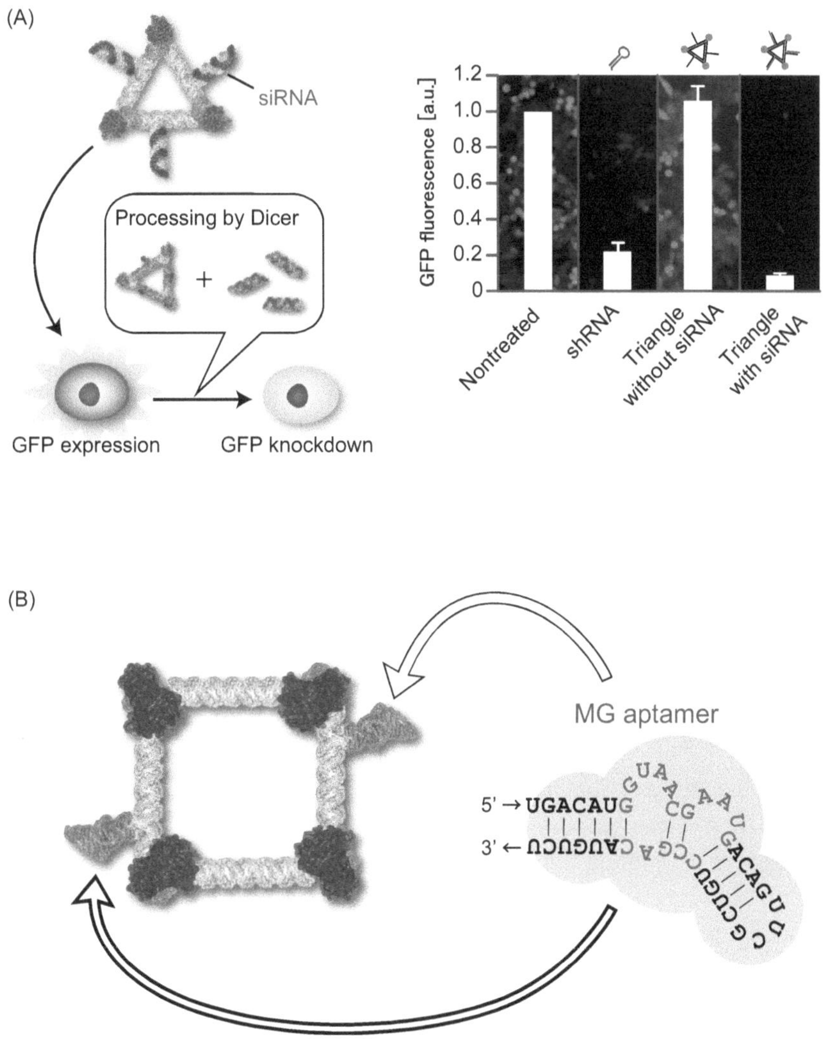

Figure 4 Introduction of functional RNA into RNP nanostructure. (A) Anti-GFP siRNA was installed onto the line regions of triangle (left). The siRNA-carrying triangle shows the effective suppression of target GFP gene expression (right graph; redrawn from ref. 31); and (B) 3D model of MG aptamer -installed L1-square (left). The 2D structure of inserted aptamer is shown (right).

siRNA-carrying triangle was transfected to human cultured cells and the expression of the target gene (GFP) was inhibited by siRNA. The efficiency of inhibition was comparable to control shRNA that is generally used for RNAi (Fig. 4A, right).

In L1-square, the RNA aptamer was installed (Fig. 4B).[39] The malachite green (MG) aptamer binds a fluorescent dye, MG and increases its fluorescence.[46] To confirm the function of the aptamer, the fluorescence was measured. The square possessing aptamer showed enhanced MG fluorescence, and the fluorescence was not affected by square-bound L1 proteins.

Other functional RNA (eg, ribozyme) or single-stranded RNA region to hybridize with DNA strand could also be introduced into the RNP scaffold.

3.2 As a Carrier for Protein

Protein molecules play various and versatile functions in organisms. Thus, if one could incorporate any protein of choice to the nanostructure, it would facilitate creation of novel functional molecule. By using RNP nanostructure, functional protein should be introduced through the fusion with the RNA binding protein of RNP module. The construction of fusion proteins is a popular technique. In that case, no chemical modification (eg, biotin for streptavidin-tag and nitrilotriacetic acid for polyhistidine-tag) is necessary.

To introduce the functional protein onto the L7-triangle, fluorescent proteins (GFP and YFP) or B1 domain of Streptococcal protein G were fused to L7Ae protein. The results by EMSA and AFM imaging showed that the fusion proteins were successfully introduced onto the triangle.[36,47] It was also confirmed that the loaded proteins kept their innate functions, the fluorescence or the binding ability to IgG.

Osada et al. applied the functional protein-carrying triangle to detect a specific cancer cell line depending on the cell surface receptor.[38] In some breast cancer cell lines, human epidermal growth factor receptor 2 (HER2) is over expressed. And, affibody$_{ZHER2:342}$ is known as a HER2 binding peptide.[48,49] We synthesized the affibody-fused L7Ae, and assembled the triangular structure with the fusion protein and fluorescent molecule-labeled RNA (Fig. 5A, left). AFM observation showed the presence of affibody at the vertices of the triangle. This complex was strewed to some human cancer cells. HER2 positive breast cancer cell line, SKBR3 was bound and labeled by the triangle, whereas other breast cancer cell lines that were HER2

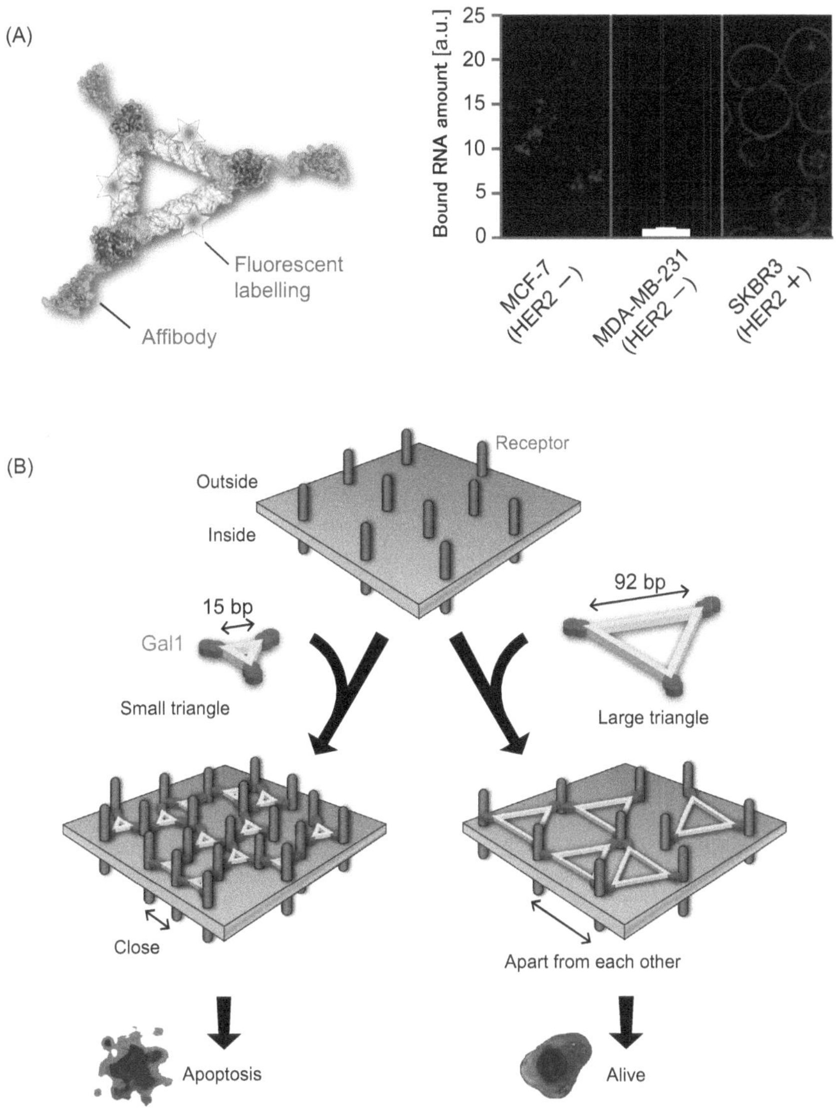

Figure 5 Functionalization of RNP nanostructure by the introduction of protein. (A) For cell type-specific recognition, affibody was introduced onto the L7-triangle (left 3D model). Triangular RNA was labelled by fluorescent dye to detect the binding to cell surface. The affibody-carrying triangle can detect specifically HER2-expressing cell (right graph; redrawn from ref. 31). (B) Control of intermolecular distance by using RNP nanostructure. Based on their size, triangular RNPs were able to control the assembling of receptor proteins on the cell-surface.

negative did not show the fluorescence (Fig. 5A, right). This cell-type specific recognition depending on the cell-surface receptors could be used for cell-type specific detection and delivery of RNA.

L7-triangle can possess three proteins since there are three RNP modules. Therefore, if three kinds of proteins are blended, trifunctional complexes should be obtained stochastically. In addition, unexpected interactions between loaded proteins should be prevented because the termini of L7Ae proteins are efferent directions of each other.

3.3 As a Size-Controllable Scaffold

It is a challenge to modify or engineer the RNA and RNP motifs, due to the unique inter/intra-molecular interactions. On the other hand, duplex region of RNA has a common A-form structure independent of the sequence. Thereby, modification such as a sequence substitution is easy in duplex region. Also, length of the line region can be easily elongated or shortened by the one pitch of A-form double-stranded helix (11 bp), retaining the steric, due to the regular structure. It implies that we can easily control distance and configuration between loaded proteins by changing the length of the duplex RNA.

Fujita et al. applied such trait of the RNP nanostructure to control receptor-mediated cellular signaling (Fig. 5B).[47] Galectin-1 (Gal1) protein is known to bind to the target receptors on cell-surface, and induce the apoptosis signal in thymocytes and T-cells, through the accumulation of target receptors. They fused L7Ae and Gal1 beta (incapable of self-dimerization), and the fusion protein was incorporated to L7-boxC/D triangle. The obtained triangular RNP was coincubated with cultured cells, and the induction of cell death by the bound RNP nanostructure was investigated. Smaller triangular RNP, which had approximately 14 nm per side, induced apoptosis efficiently (Fig. 5B, lower left). This indicates the accumulation and interaction of receptors facilitated by Gal1 beta-including triangle RNP. Notably, increasing the length of the RNP triangle side was inversely proportional to the induction of cell-death. Finally, larger triangular RNP, in which the distance between loaded proteins was more than 30 nm, did not induce cell death (Fig. 5B, lower right). It is considered that the large triangle RNP captured the receptors but kept them apart from each other so that it did not trigger cell death. These results show that the RNP nanostructure is a suitable material to precisely control the positioning of molecules at the nanometer level and regulate their interactions to alter the cellular fate.

3.4 In Cell Applications

Both components of RNP nanostructures, that is, RNA and protein, can be synthesized inside of the cell by introducing genes that code for them. Therefore, RNP nanostructures should be able to be constructed and work inside the cell. To do so, it is necessary to fold RNA in isothermal conditions that reflect cellular conditions, instead of the annealing treatment that is generally performed in DNA nanotechnology to make the complementary-strand hybridization efficiently. Indeed, some RNA and RNP nanostructures were formed in isothermal conditions.[25,26,39]

Control of intermolecular interactions is also important to cell functions. Molecular scaffold is often employed for the purpose.[50–52] The scaffold works through concentration, colocalization and arrangement of the target molecules. For example, a scaffold protein, Ste5 is known to play an important role in MAPK cascade signaling pathway.[53] In metabolic pathways, involved enzymes are often colocalized onto a scaffold to promote sequential reactions.[54,55] Telomerase RNA serves as a scaffold for the protein subunits in yeast.[56]

Referring to natural scaffolds, some synthetic protein scaffolds have been reported.[57,58] For example, Ryu et al. succeeded in designing the scaffold protein with PDZ domain to regulate cell signaling through a MAPK cascade in yeast and mammalian cells.[59] Synthetic DNA scaffold was also reported.[60] In this research, enzymes were arranged onto a DNA strand using zinc-finger proteins that are sequence-dependent DNA-binding proteins.

RNA nanostructure can also be used to assemble proteins. In fact, Silver and coworkers have used RNA scaffolds that can work inside cells.[61,62] They designed some RNA nanostructures and placed enzyme proteins onto the scaffold using various RNP motifs such as PP7, MS2, Rev and N-peptide and their binding partner RNA. In living *E. coli*, the RNA scaffolds increased the desired metabolic flux effectively by the colocalization of enzymes involved in the metabolic pathway.

The position, distance and orientation between the molecules are critical in the control of the intermolecular interaction. As described earlier, the size and shape of RNA and RNP nanostructures can be modified easily. By the use of RNP motifs and RNP nanostructures, highly sophisticated regulations of a biochemical pathway would be realized, for example, accurate regulation of intermolecular distance and the functional regulation depending on the presence of specific proteins. In addition to such advanced regulations, the application in eukaryotic cells including human cells and the targeting of endogenous proteins are the future challenges.

4. RNP MOTIFS AS A BIOLOGICAL TOOL

The utility of RNP interacting motifs is not limited only to be used as building blocks in constructing nanostructures. Various applications have been proposed.[63] In our group, protein-responsive "RNA switches" were created with the RNP motifs to control gene expression (Fig. 6).

To design and construct an "off switch" (the specific RNP interaction represses gene expression; Fig. 6A), the target L7Ae protein-binding box C/D RNA was inserted simply at the 5'-UTR or ORF region of the reporter mRNA.[64] In the absence of L7Ae protein, the mRNA can be translated as usual ("on" state; Fig. 6A, left). However, in the presence of L7Ae, the protein binds to the mRNA and inhibits the function of the ribosome to initiate or proceed translation ("off" state; Fig. 6A, right). The inhibition efficiency is tuned by the kind of RNP module, insertion region and the number of inserted modules. In addition to "off switches", protein-responsive RNA "on switches" (the RNP interaction activate gene expression) have been designed and shown to be functional in live mammalian cells (Fig. 6B).[65,66] In these systems, RNP modules serve to bring target protein onto the specific region of RNA and interfere with the natural systems by the steric hindrance. Intermolecular distance and orientation of each molecule are important for the effective control of steric hindrance or molecular interaction. Molecular design tools based on 3D structures of RNA/RNP will help to design effective switches. Indeed, Kashida et al. showed that the estimation from 3D molecular modeling of RNA-switches was consistent with the biological effect in cell.[67] Thus, RNP module will promote the application by the structure-based design.

Thus, the same RNA and RNP motifs can be used to design both nanostructures and genetic switches. There are many RNP motifs in naturally occurring molecules, although we tested and used only a few kinds of RNP motifs.[68,69] Moreover, new synthetic RNP motifs may be obtained by the technique known as "SELEX" or "in vitro selection."[70,71]

5. CONCLUSION AND FUTURE PERSPECTIVES

Variety of nanostructures can be made of RNA and RNP motifs. These nanostructures can be applied in nanomedicine field because they can easily carry functional molecules. In addition, gene expression and cell

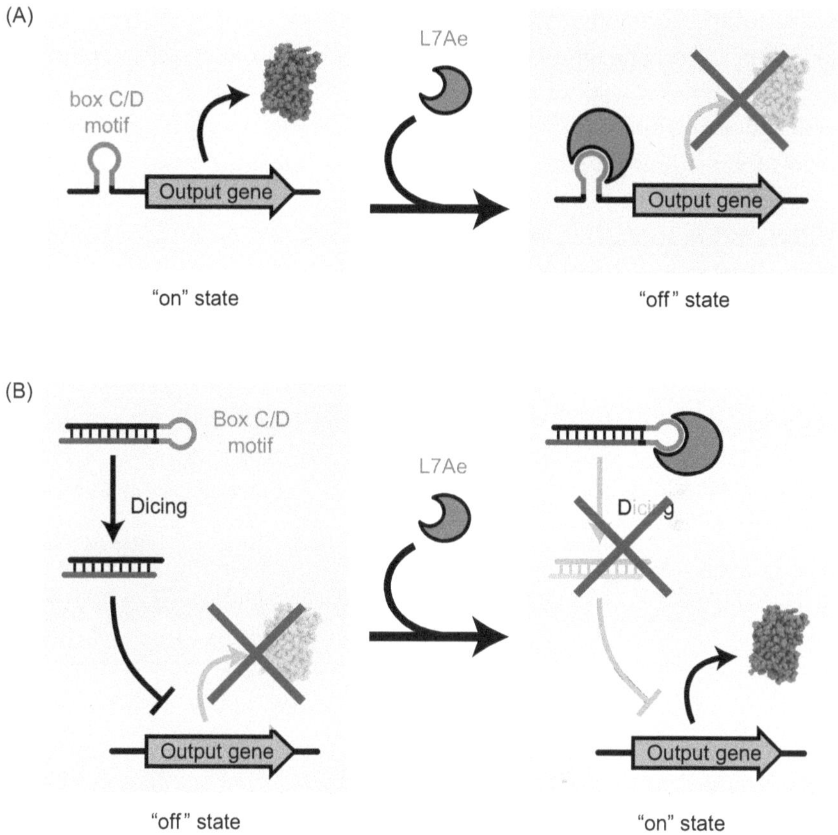

Figure 6 RNP switch to control gene expression. (A) "OFF switch" by translational inhibition. In the absence of L7Ae protein, the mRNA can be translated by ribosome and the output gene is expressed ("on" state; left). However, in the presence of L7Ae, it binds to box C/D motif on the mRNA and impedes the translation ("off" state; right).(B) "ON switch" employing RNA interference pathway. Here, RNP motif is inserted to a premicro RNA (pre-miRNA). In the absence of L7Ae protein, the pre-miRNA is normally processed into micro RNA (miRNA) by Dicer. The miRNA represses the expression of reporter mRNA ("off" state; left). In contrast, L7Ae binds to pre-miRNA and impedes the processing by Dicer in the presence of L7Ae. Therefore, reporter mRNA is expressed ("on" state; right).

fate regulation systems including scaffolds and translational switches have been developed using RNA and RNP motifs. Thus, RNA and RNP themselves are useful building blocks and valid tools in the fields of biotechnology. At present, reported RNA and RNP-based nanostructures have comparatively simple shapes. And only a few kinds of motifs have been exploited for the construction of these nanostructures. If we get more various motifs for design, we maybe able to create more various molecules that have 3D complexities and/or the catalytic abilities. Such attempts to design complex nanostructures may also serve to elucidate the building principles of natural occurring RNA or RNP molecules, and give us hints to construct sophisticated RNP nanomachines, such as ribosome.

ACKNOWLEDGMENTS

We thank Dr Callum Parr (Kyoto University) for proofreading and insightful comments. This work was supported by a Grant-in-Aid for Scientific Research on Innovative Areas "Molecular Robotics" (No. 24104002) and Grant-in-Aid for Scientific Research (S) (No. 15H05722) from the Ministry of Education, Culture, Sport, Science, and Technology, Japan.

REFERENCES

1. Seeman NC. Nucleic acid junctions and lattices. *J Theor Biol.* 1982;99:237–247.
2. Chen JH, Seeman NC. Synthesis from DNA of a molecule with the connectivity of a cube. *Nature.* 1991;350:631–633.
3. Rothemund PW. Folding DNA to create nanoscale shapes and patterns. *Nature.* 2006;440:297–302.
4. Endo M, Yang Y, Sugiyama H. DNA origami technology for biomaterials applications. *Biomater Sci-UK.* 2013;1:347–360.
5. Guo P. The emerging field of RNA nanotechnology. *Nat Nanotechnol.* 2010;5: 833–842.
6. Shukla GC, Haque F, Tor Y, et al. A boost for the emerging field of RNA nanotechnology. *ACS Nano.* 2011;5:3405–3418.
7. Grabow W, Jaeger L. RNA modularity for synthetic biology. *F1000Prime Rep.* 2013;5:46.
8. Klein DJ, Schmeing TM, Moore PB, Steitz TA. The kink-turn: a new RNA secondary structure motif. *EMBO J.* 2001;20:4214–4221.
9. Ennifar E, Walter P, Ehresmann B, Ehresmann C, Dumas P. Crystal structures of coaxially stacked kissing complexes of the HIV-1 RNA dimerization initiation site. *Nat Struct Biol.* 2001;8:1064–1068.
10. Schuwirth BS, Borovinskaya MA, Hau CW, et al. Structures of the bacterial ribosome at 3.5 Å resolution. *Science.* 2005;310:827–834.
11. Hingerty B, Brown RS, Jack A. Further refinement of the structure of yeast tRNAPhe. *J Mol Biol.* 1978;124:523–534.
12. Dibrov SM, Johnston-Cox H, Weng YH, Hermann T. Functional architecture of HCV IRES domain II stabilized by divalent metal ions in the crystal and in solution. *Angew Chem Int Ed Engl.* 2007;46:226–229.
13. Ding F, Lu C, Zhao W, et al. Structure and assembly of the essential RNA ring component of a viral DNA packaging motor. *Proc Nat Acad Sci USA.* 2011;108: 7357–7362.
14. Lee AJ, Crothers DM. The solution structure of an RNA loop–loop complex: the ColE1 inverted loop sequence. *Structure.* 1998;6:993–1005.
15. Cate JH, Gooding AR, Podell E, et al. Crystal structure of a group I ribozyme domain: principles of RNA packing. *Science.* 1996;273:1678–1685.
16. Chworos A, Severcan I, Koyfman AY, et al. Building programmable jigsaw puzzles with RNA. *Science.* 2004;306:2068–2072.
17. Dibrov SM, McLean J, Parsons J, Hermann T. Self-assembling RNA square. *Proc Natl Acad Sci USA.* 2011;108:6405–6408.

18. Severcan I, Geary C, Verzemnieks E, Chworos A, Jaeger L. Square-shaped RNA particles from different RNA folds. *Nano Lett.* 2009;9:1270–1277.
19. Khaled A, Guo S, Li F, Guo P. Controllable self-assembly of nanoparticles for specific delivery of multiple therapeutic molecules to cancer cells using RNA nanotechnology. *Nano Lett.* 2005;5:1797–1808.
20. Khisamutdinov EF, Li H, Jasinski DL, et al. Enhancing immunomodulation on innate immunity by shape transition among RNA triangle, square and pentagon nanovehicles. *Nucleic Acids Res.* 2014;42:9996–10004.
21. Grabow WW, Zakrevsky P, Afonin KA, et al. Self-assembling RNA. nanorings based on RNAI/II inverse kissing complexes. *Nano Lett.* 2011;11:878–887.
22. Severcan I, Geary C, Verzemnieks E, Chworos A, Jaeger L. A polyhedron made of tRNAs. *Nat Chem.* 2010;2:772–779.
23. Hao C, Li X, Tian C, et al. Construction of RNA nanocages by re-engineering the packaging RNA of Phi29 bacteriophage. *Nat Commun.* 2014;5:3890.
24. Nasalean L, Baudrey S, Leontis NB, Jaeger L. Controlling RNA self-assembly to form filaments. *Nucleic Acids Res.* 2006;34:1381–1392.
25. Geary C, Rothemund PW, Andersen ES. A single-stranded architecture for cotranscriptional folding of RNA nanostructures. *Science.* 2014;345:799–804.
26. Afonin KA, Bindewald E, Yaghoubian AJ, et al. In vitro assembly of cubic RNA-based scaffolds designed in silico. *Nat Nanotechnol.* 2010;5:676–682.
27. Yu J, Liu Z, Jiang W, Wang G, Mao C. De novo design of an RNA tile that self-assembles into a homo-octameric nanoprism. *Nat Commun.* 2015;6:5724.
28. Hamada S, Murata S. Substrate-assisted assembly of interconnected single-duplex DNA nanostructures. *Angew Chem Int Ed Engl.* 2009;48:6820–6823.
29. Afonin KA, Viard M, Koyfman AY, et al. Multifunctional RNA nanoparticles. *Nano Lett.* 2014;14:5662–5671.
30. Khisamutdinov EF, Jasinski DL, Guo P. RNA as a boiling-resistant anionic polymer material to build robust structures with defined shape and stoichiometry. *ACS Nano.* 2014;8:4771–4781.
31. Jasinski DL, Khisamutdinov EF, Lyubchenko YL, Guo P. Physicochemically tunable polyfunctionalized RNA square architecture with fluorogenic and ribozymatic properties. *ACS Nano.* 2014;8:7620–7629.
32. Guo S, Huang F, Guo P. Construction of folate-conjugated pRNA of bacteriophage phi29 DNA packaging motor for delivery of chimeric siRNA to nasopharyngeal carcinoma cells. *Gene Ther.* 2006;3:814–820.
33. Yusupov MM, Yusupova GZ, Baucom A, et al. Crystal structure of the ribosome at 5.5 Å resolution. *Science.* 2001;292:883–896.
34. Selmer M, Dunham CM, Murphy FV t, et al. Structure of the 70S ribosome complexed with mRNA and tRNA. *Science.* 2006;313:1935–1942.
35. Ben-Shem A, Jenner L, Yusupova G, Yusupov M. Crystal structure of the eukaryotic ribosome. *Science.* 2010;330:1203–1209.
36. Ohno H, Kobayashi T, Kabata R, et al. Synthetic RNA-protein complex shaped like an equilateral triangle. *Nat Nanotechnol.* 2011;6:116–120.
37. Turner B, Melcher SE, Wilson TJ, Norman DG, Lilley DM. Induced fit of RNA on binding the L7Ae protein to the kink-turn motif. *RNA.* 2005;11: 1192–1200.
38. Osada E, Suzuki Y, Hidaka K, et al. Engineering RNA-protein complexes with different shapes for imaging and therapeutic applications. *ACS Nano.* 2014;8: 8130–8140.
39. Ohno H, Inoue T. Designed Regular Tetragon-Shaped RNA-Protein Complexes with Ribosomal Protein L1 for Bionanotechnology and Synthetic Biology. *ACS Nano.* 2015;9:4950–4956.

40. Nevskaya N, Tishchenko S, Gabdoulkhakov A, et al. Ribosomal protein L1 recognizes the same specific structural motif in its target sites on the autoregulatory mRNA and 23S rRNA. *Nucleic Acids Res.* 2005;33:478–485.
41. Kohrer C, Mayer C, Neumair O, Grobner P, Piendl W. Interaction of ribosomal L1 proteins from mesophilic and thermophilic Archaea and Bacteria with specific L1-binding sites on 23S rRNA and mRNA. *Eur J Biochem.* 1998;256: 97–105.
42. Flierl U, Nero TL, Lim B, et al. Phosphorothioate backbone modifications of nucleotide-based drugs are potent platelet activators. *J Exp Med.* 2015;212:129–137.
43. Layzer JM, McCaffrey AP, Tanner AK, et al. In vivo activity of nuclease-resistant siRNAs. *RNA.* 2004;10:766–771.
44. GaglioneM. Messere A. Recent progress in chemically modified siRNAs. *Mini Rev Med Chem.* 2010;10:578–595.
45. Ohuchi SJ, Sagawa F, Ohno H, Inoue T. A purification method for a molecular complex in which a scaffold molecule is fully loaded with heterogeneous molecules. *PLoS One.* 2015;10:e0120576.
46. Baugh C, Grate D, Wilson C. 2.8 Å crystal structure of the malachite green aptamer. *J Mol Biol.* 2000;301:117–128.
47. Fujita Y, Furushima R, Ohno H, Sagawa F, Inoue T. Cell-surface receptor control that depends on the size of a synthetic equilateral-triangular RNA-protein complex. *Sci Rep.* 2014;4:6422.
48. Orlova A, Magnusson M, Eriksson TL, et al. Tumor imaging using a picomolar affinity HER2 binding affibody molecule. *Cancer Res.* 2006;66:4339–4348.
49. Eigenbrot C, Ultsch M, Dubnovitsky A, Abrahmsen L, Hard T. Structural basis for high-affinity HER2 receptor binding by an engineered protein. *ProcNat Acad Sci USA.* 2010;107:15039–15044.
50. Burack WR, Shaw AS. Signal transduction: hanging on a scaffold. *Curr Opin Cell Biol.* 2000;12:211–216.
51. GoodMC. Zalatan JG, Lim WA. Scaffold proteins: hubs for controlling the flow of cellular information. *Science.* 2011;332:680–686.
52. Chen R, Chen Q, Kim H, et al. Biomolecular scaffolds for enhanced signaling and catalytic efficiency. *Curr Opin Biotechnol.* 2014;28:59–68.
53. Malleshaiah MK, Shahrezaei V, Swain PS, Michnick SW. The scaffold protein Ste5 directly controls a switch-like mating decision in yeast. *Nature.* 2010;465: 101–105.
54. Morais S, Heyman A, Barak Y, et al. Enhanced cellulose degradation by nano-complexed enzymes: synergism between a scaffold-linked exoglucanase and a free endoglucanase. *J Biotechnol.* 2010;147:205–211.
55. You C, Myung S, Zhang YH. Facilitated substrate channeling in a self-assembled trifunctional enzyme complex. *Angew Chem Int Ed Engl.* 2012;51:8787–8879.
56. Zappulla DC, Cech TR. Yeast telomerase RNA: a flexible scaffold for protein subunits. *Proc Nat Acad Sci USA.* 2004;101:10024–10029.
57. Park SH, Zarrinpar A, Lim WA. Rewiring MAP kinase pathways using alternative scaffold assembly mechanisms. *Science.* 2003;299:1061–1064.
58. Dueber JE, Wu GC, Malmirchegini GR, et al. Synthetic protein scaffolds provide modular control over metabolic flux. *Nat Biotechnol.* 2009;27:753–759.
59. Ryu J, Park SH. Simple synthetic protein scaffolds can create adjustable artificial MAPK circuits in yeast and mammalian cells. *Sci Signal.* 2015;8:ra66.
60. Conrado RJ, Wu GC, Boock JT, et al. DNA-guided assembly of biosynthetic pathways promotes improved catalytic efficiency. *Nucleic Acids Res.* 2012;40:1879–1889.
61. Delebecque CJ, Lindner AB, Silver PA, Aldaye FA. Organization of intracellular reactions with rationally designed RNA assemblies. *Science.* 2011;333:470–474.

62. Sachdeva G, Garg A, Godding D, Way JC, Silver PA. In vivo colocalization of enzymes on RNA scaffolds increases metabolic production in a geometrically dependent manner. *Nucleic Acids Res.* 2014;42:9493–9503.
63. Mackay JP, Font J, Segal DJ. The prospects for designer single-stranded RNA-binding proteins. *Nat Struct Mol Biol.* 2011;18:256–261.
64. Saito H, Kobayashi T, Hara T, et al. Synthetic translational regulation by an L7Ae-kink-turn RNP switch. *Nat Chem Biol.* 2010;6:71–78.
65. Saito H, Fujita Y, Kashida S, Hayashi K, Inoue T. Synthetic human cell fate regulation by protein-driven RNA switches. *Nat Commun.* 2011;2:160.
66. Endo K, Hayashi K, Inoue T, Saito H. A versatile cis-acting inverter module for synthetic translational switches. *Nat Commun.* 2013;4:2393.
67. Kashida S, Inoue T, Saito H. Three-dimensionally designed protein-responsive RNA devices for cell signaling regulation. *Nucleic Acids Res.* 2012;40:9369–9378.
68. Saito H, Inoue T. RNA and RNP as new molecular parts in synthetic biology. *J Biotechnol.* 2007;132:1–7.
69. Saito H, Inoue T. Synthetic biology with RNA motifs. *Int J Biochem Cell Biol.* 2009;41:398–404.
70. Tuerk C, Gold L. Systematic evolution of ligands by exponential enrichment: RNA ligands to bacteriophage T4 DNA polymerase. *Science.* 1990;249:505–510.
71. Ellington AD, Szostak JW. In vitro selection of RNA molecules that bind specific ligands. *Nature.* 1990;346:818–822.

INDEX

Q

R

Zinc-finger proteins (ZFPs) 138, 179

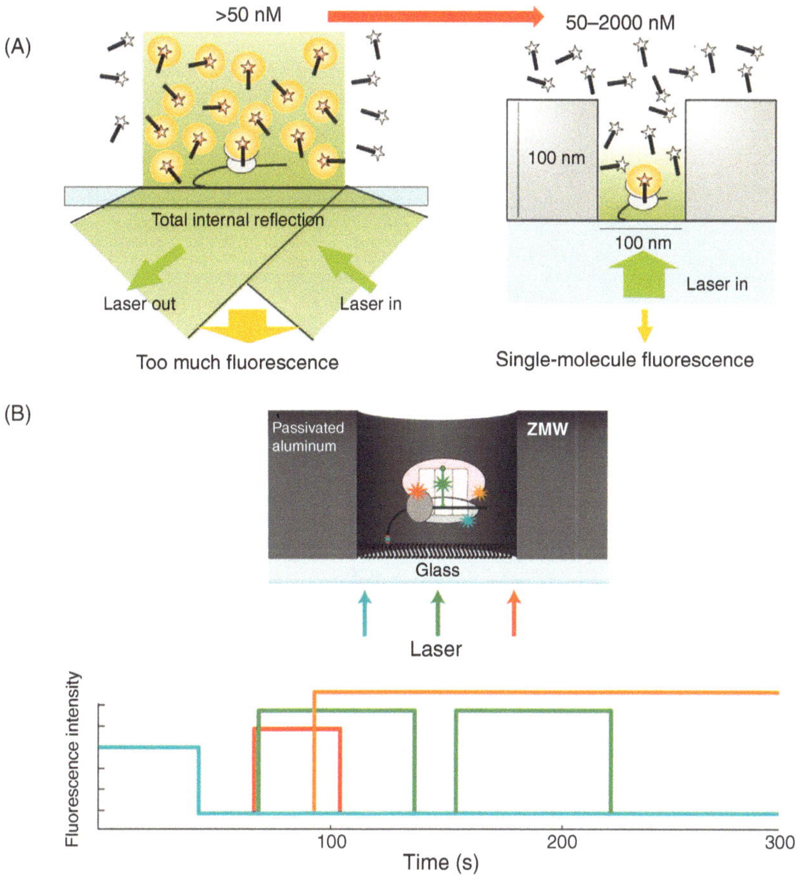

Albert Tsai *et al.*, Figure 1 Zero-mode waveguide (ZMW) allows translation to be tracked using single-molecule fluorescence at near-physiological conditions. (A) Conventional single-molecule setup using total internal reflection fluorescence (TIRF) microscopy illuminates about 150 nm into the solution containing freely diffusing molecules labeled with fluorophores. This limits the concentration of labeled ligands in solution down to 50 nM or less at normal imaging speed of 10 times per second. ZMWs concentrate the laser excitation down to only 20–30 nm from the bottom of the well, allowing more than 1 μM of labeled molecules to be present in solution. (B) In order to make the ZMW structure compatible with biological molecules, the aluminum surface was passivated and the exposed glass surface at the bottom of the ZMW was treated with biotinylated polyethylene glycol (PEG). Biotinylated molecules can be immobilized inside the ZMW through the use of streptavidin or neutravidin. The increased concentration and the multiplexing ability of the ZMW microscope to observe several different color channels simultaneously allowed the dynamics of ribosomes translating on complex mRNA sequences to be tracked with a full complement of translation factors and tRNA at near-physiological concentrations.

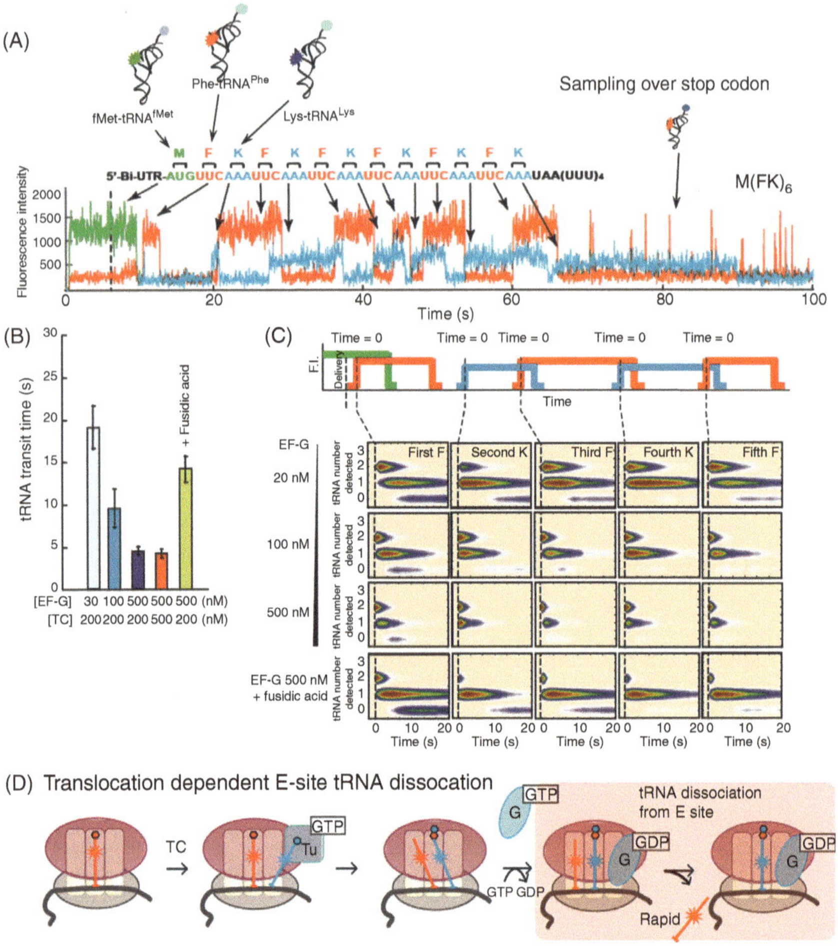

Albert Tsai *et al.*, Figure 2 Applying the ZMW technology to study translation: tRNA transit through the ribosome during elongation. (A) Tracking tRNA transit through the ribosome during elongation at near-physiological concentrations of fluorescently labeled tRNAs (tRNA^{fMet} with Cy3, tRNA^{Phe} with Cy5, and tRNA^{Lys} with Cy2, where each tRNA is indicated using their one-letter amino acid code in the sequence of pulses) was possible through imaging ribosome complexes in ZMW wells. Translating an mRNA coding for six repeating Phe and Lys codons gave the correct sequence of fluorescent pulses (alternating F and K pulses), verifying that ribosomes translate normally within ZMWs. Many ribosomes translated the full message and begin to show short tRNA sampling pulses over the final stop codon as no termination and ribosome release factors were available to terminate translation. (B) Working at high concentrations of tRNA and EF-G, translation elongation was efficient and processive. With 500 nM of labeled tRNAs in solution, translation could proceed at a rate of less than 5 s per codon under single-molecule imaging conditions. (C) Tracking tRNA occupancy on the ribosome during translation elongation revealed that tRNA departure from the E site is dependent on translocation. As translocation rates increased with increasing EF-G concentration, the time that two tRNAs occupied the ribosome decreased. With fusidic acid blocking EF-G dissociation and thus A-site tRNA arrival, ribosomes still quickly evolved into single-tRNA occupancy. E-site tRNA departure therefore occurs stochastically after translocation, without clear correlation to the arrival of a new tRNA in the A site. (D) This result supports the model of E-site tRNA departure where translocation is the gating event immediately prior to tRNA dissociation. As soon as the ribosome translocates, the E-site tRNA rapidly and stochastically departs without the need for another tRNA to accommodate in the A site.

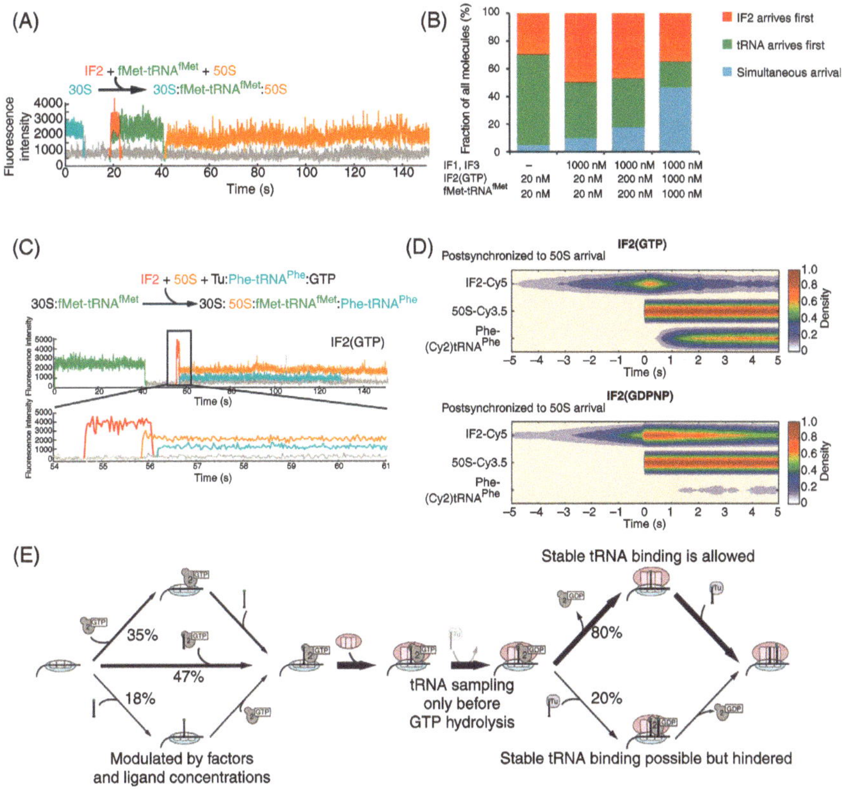

Albert Tsai *et al.*, Figure 3 Tracking the assembly of a translationally competent ribosome complex through initiation. (A) With all of the essential components for prokaryotic translation initiation labeled, the entire initiation pathway from 30S PIC formation to subunit joining could be monitored in the ZMW. (B) All possible sequences of IF2 and initiator tRNA arrival was observed in the formation of functional 30S PICs. The preferred pathway leading to the assembly of a functional 30S PIC varied depending on the presence of other factors (eg, IF1 and IF3) and the concentrations of IF2 and initiator tRNA. At high concentrations of both, which is closest to the condition in vivo, simultaneous arrival of both molecules as a complex was the predominant mode of 30S PIC formation. (C) Tracking the ribosome's transition from initiation into elongation revealed the role of IF2 in controlling the arrival of the first elongator tRNA. (D) Postsynchronizing to the arrival of large 50S subunit (aligning all events to have 50S arrival to be $t = 0$) showed that IF2 departs within 2 s after subunit joining, followed by the arrival of the first elongator tRNA. When IF2 cannot hydrolyze GTP, as was the case with the nonhydrolyzable GTP analog GDPNP, it remained on the ribosome after subunit joining and blocked the arrival of the first elongator tRNA. Therefore IF2 departure subsequent to GTP hydrolysis is the final gating event for elongation to begin. (E) Investigating translation initiation at near-physiological conditions using the ZMW highlighted the complexity and the heterogeneous pathways that could potentially serve as points of regulation before the ribosome fully commits to translating a full protein. Nonessential factors, such as IF1 and IF3 likely modify the flux through each pathway, leading to further layers of regulation and fine tuning.

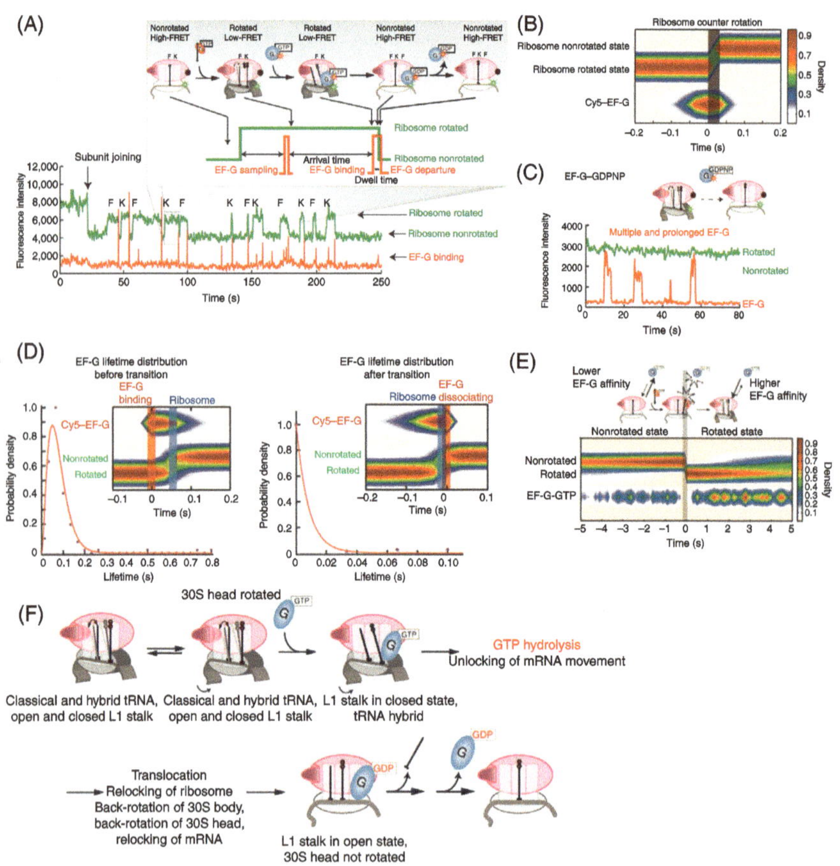

Albert Tsai *et al.*, Figure 4 Dissecting the mechanism of EF-G catalyzed translocation. (A) With sufficiently high concentrations of Cy5-labeled EF-G (red), clear binding events could be seen coupled with intersubunit conformation change from the rotated state back to the nonrotated state, which is coupled to translocation. (B) Correlating the EF-G binding events to the exact time of intersubunit conformation change showed that EF-G binds approximately 70 ms prior to intersubunit counter rotation and dissociates rapidly afterward in about 30 ms. (C) When EF-G cannot hydrolyze GTP, as was the case with the nonhydrolyzable GDPNP, prolonged EF-G binding events that failed to drive translocation became frequent. (D) Fitting the dwell time of EF-G prior to intersubunit rotation/translocation yielded a Poisson distributed lifetime, indicating that there are multiple rate-limiting steps. GTP hydrolysis is likely one of them. Posttranslocation, the dwell time was exponential, suggesting only a single rate-limiting step is present, which is EF-G dissociation from the ribosome. (E) The affinity of the ribosome for EF-G changed depending on its rotational state. Upon transition from the nonrotated to the rotated state, EF-G sampling frequency increased as seen in the higher density in the EF-G-GTP row of the plot. (F) EF-G catalyzes translocation first by binding to the ribosome in the rotated state and then guiding the internal conformational fluctuations of the ribosome to the correct configuration for translocation. GTP hydrolysis then drives translocation and both the ribosome and EF-G undergo major conformational changes. Once translocation has completed, EF-G-GDP departs from the ribosome rapidly due to its low affinity for remaining bound to the ribosome.

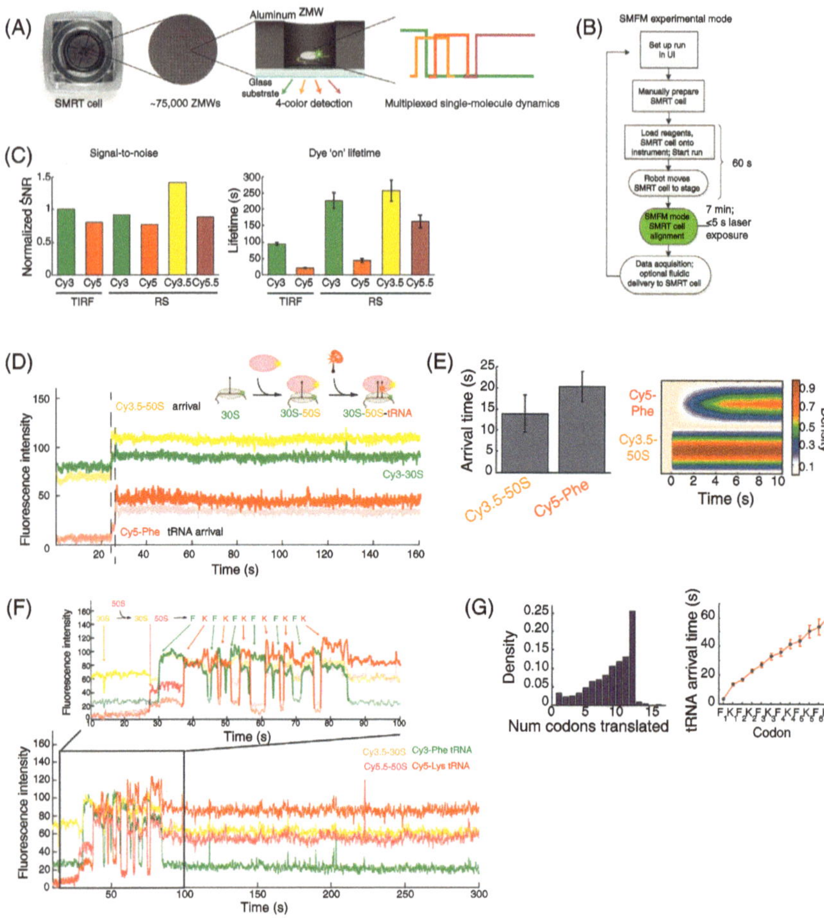

Albert Tsai *et al.*, Figure 5 A modified RS platform for use as a general single-molecule fluorescence microscope. (A) The commercial RS instrument uses the SMRT cell containing 75,000–150,000 ZMW wells, all of which can be imaged simultaneously in four different color channels. (B) The modified workflow for single-molecule fluorescence imaging is flexible toward different experimental protocols involving live deliveries of solutions into the chip during imaging and minimizes unduly exposing the ZMW chip to laser illumination prior to the experiment. (C) The four color channels, corresponding to the commercially Cy3, Cy3.5, Cy5, and Cy5.5 dyes exhibited comparable signal-to-noise ratios compared to a conventional TIRF microscope. The oxygen-free environment inside the imaging chamber significantly improved the photobleaching lifetimes of the dyes compared to an unsealed microscope. (D) Repeating the initiation experiments of Tsai et al.[83] on the RS yielded comparable results. (E) The times required for subunit joining and elongator tRNA arrival compared favorably with the original experiments. (F) Repeating the tRNA transit experiment of Uemura et al.[54] with more than 3000 observed molecule was possible with the high throughput of the RS platform. The four-color capability additionally allowed both ribosomal subunits to be tracked. (G) The high concentration of labeled tRNAs used ensure the most molecules translated the full 12-codon mRNA, with fast translation rates for single-molecule experiments on the order of a few seconds per codon.

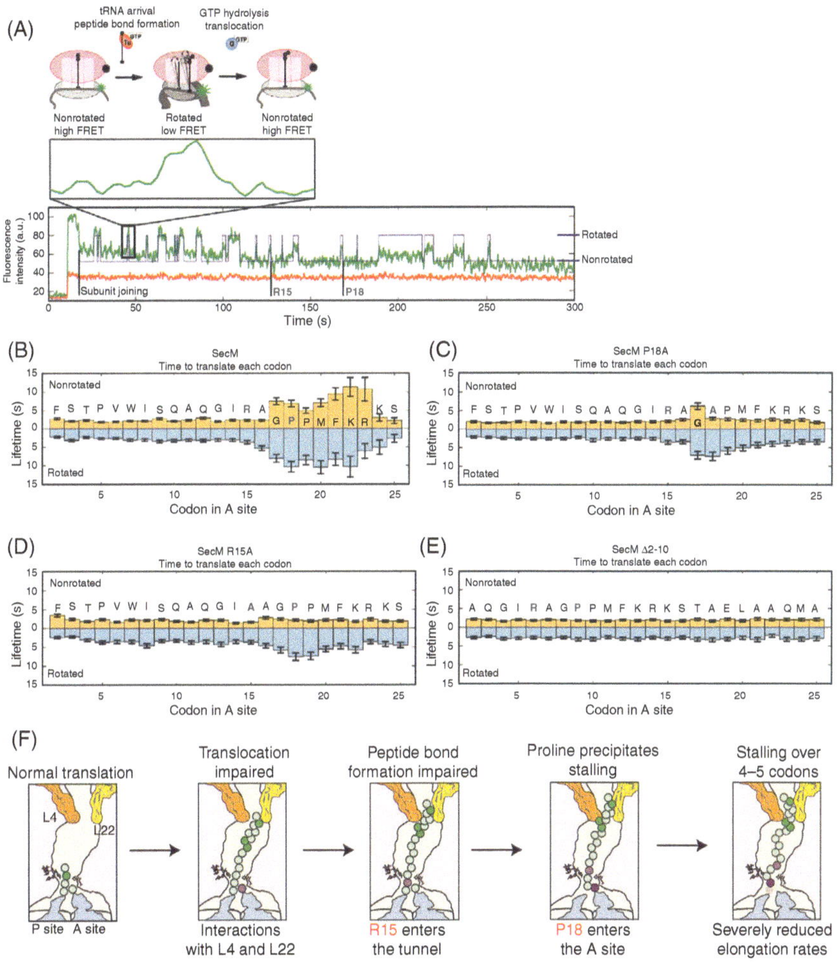

Albert Tsai *et al.*, Figure 6 Probing interactions between the nascent peptide and the exit tunnel critical in translational stalling. (A) Tracking the intersubunit conformation as the ribosome is translating an mRNA coding for the SecM stall sequence revealed that the stall sequence induces slowdowns over specific locations on the mRNA. (B) The wildtype SecM sequence induced a general slowdown in translocation as shown in the longer rotated state lifetimes after the first critical Arg15 has entered the exit tunnel. The nonrotated lifetimes also increased over codon 17 and 2 codons after the final Pro18, suggesting difficulties in peptide bond formation. (C) Mutating Pro18 to Ala removed all slowdowns in the nonrotated state lifetimes after the 18th codon and no stalling ensued. (D) Changing Arg15 to Ala removed all inhibitions in the rate of peptide bond formation while some decrease in the translocation rate could still be seen. No stalling was observed. (E) Removing the first nine amino acids of the stall sequence, while retaining the critical Arg and Pro, returned translation dynamics completely back to that comparable with nonstalling mRNA sequences. (F) The specific effects of each element on the minimal stall sequence on ribosome dynamics show that stalling is a dynamic event involving several precisely timed events in a cascade. The amino acids prior to the critical Arg15 induces peptide compaction in the tunnel to reduce translocation rate, allowing Arg15 to interact with the tunnel. Arg15 then remodels the PTC via a network of interactions starting from within the exit tunnel, slowing peptide bond formation. Pro18 then precipitates stalling through its unique structural property, locking the nascent peptide in conformations that severely reduces peptidyl transfer rates and translocation rates.

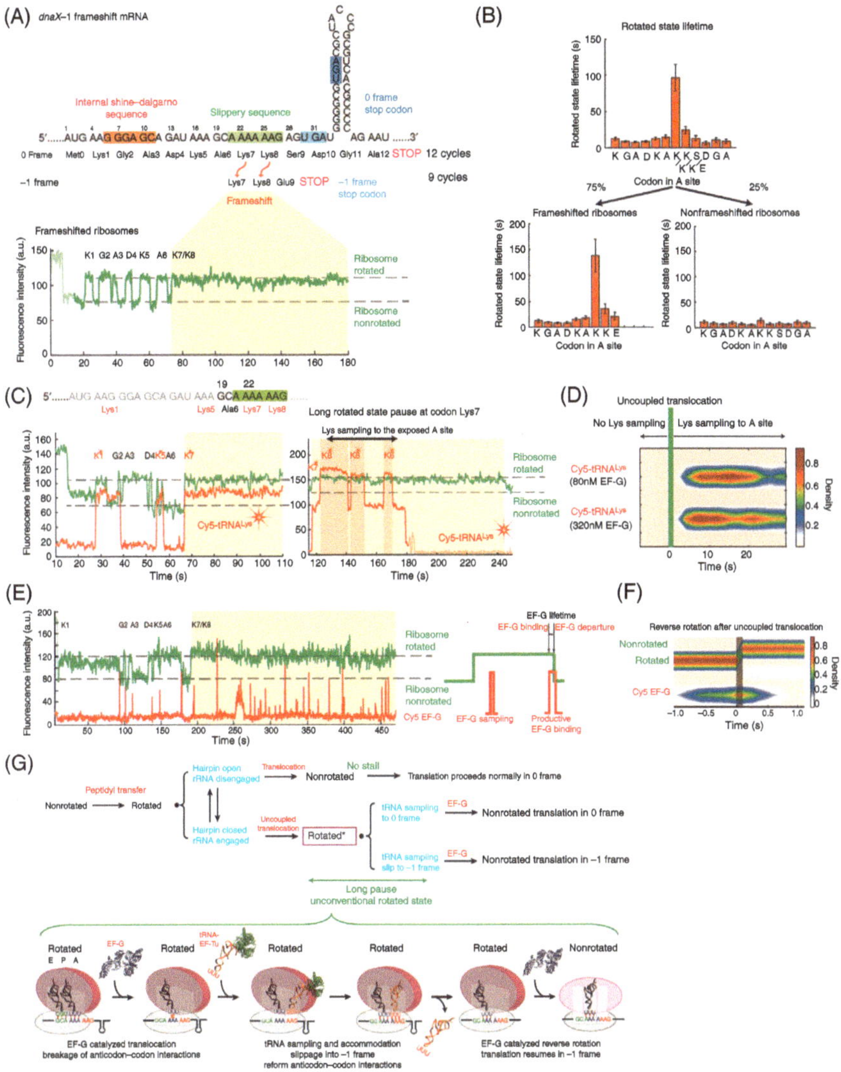

Albert Tsai *et al.*, Figure 7 The mechanism of −1 frameshifting involves coordination between tRNA, EF-G, and intersubunit conformational changes. (A) The *dnaX* sequence contains all canonical elements present in sequences that induce −1 frameshifting: an optional internal Shine–Dalgarno sequence, an ambiguous slippery sequence (X XXY YYZ) and a hairpin afterward. The specific sequence used has an open reading frame of 9 codons for frameshifted ribosomes and 12 codons for nonframeshifted ribosomes, so the fraction of frameshifted ribosomes could be separated based on how many codons they translated. (B) The native *dnaX* sequence induced a slowdown in the rotated state specifically over codon 7 over the slippery sequence. While this pause was the hallmark of −1 frameshifting occurring for 75% of the time, ribosomes that did not frameshift showed no signs of pausing. (C) Monitoring tRNA binding dynamics over the slippery sequence revealed prolonged tRNA binding to the A site, perhaps after translocation with the ribosome in a noncanonical rotated state. These sampling events likely redefine the reading frame over the slippery sequence, allowing the ribosome to frameshift because the noncanonical state has a loosened grip on the tRNAs and the mRNA. (D) The time until the appearance of tRNA sampling was directly related to the concentration of EF-G, implying that while the ribosome does not return to its canonical nonrotated state, tRNA can sample posttranslocation. (E) EF-G was critical in resolving the noncanonical rotated state, where multiple EF-G sampling events usually preceded the final binding event that successfully counter rotated the subunit. (F) The final binding event counter rotated the subunit, but the EF-G resided for a prolonged period of time compared to normal translocation, indicating that a significant energy barrier is present. (G) The mechanism of −1 frameshifting involves several branch points along the way: first the ribosome engages the 3'-hairpin and optionally the 5'-internal Shine–Dalgarno sequence in prokaryotes. An uncoupled translocation then leaves the ribosome in a noncanonical rotated state, allowing tRNAs to sample in the A site. Prolonged A-site sampling over the slippery sequence leads to a redefinition of the reading frame, inducing a −1 frameshift. Finally, an EF-G resolves the noncanonical state, returning the ribosome to normal translation.

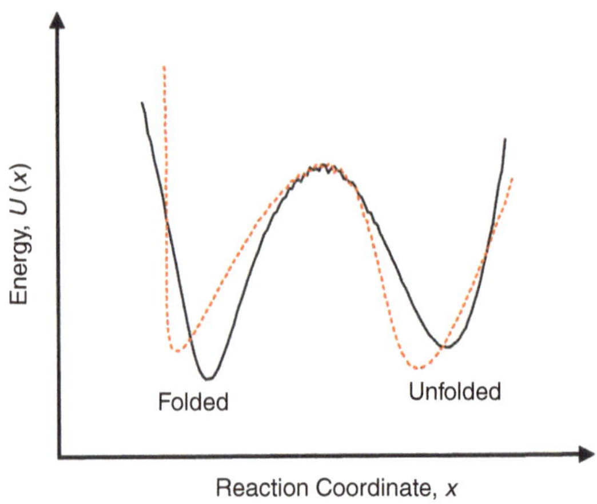

Koen Visscher, Figure 5 1D energy landscape with a kinetic barrier separating the unfolded and folded states. Application of force tilts the landscape (red dashed) so as to lower the barrier from the folded state promoting escape to the unfolded state.

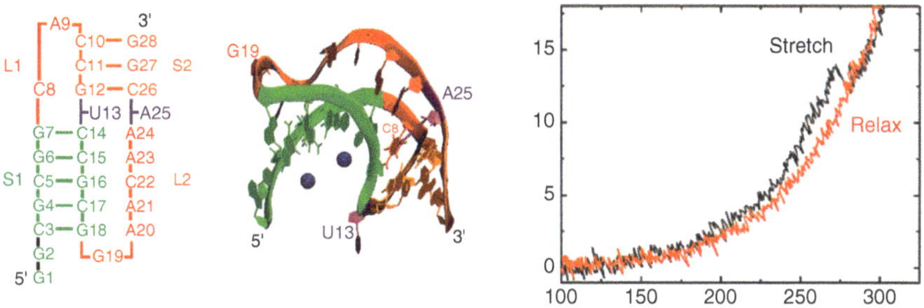

Koen Visscher, Figure 6 BWYV pseudoknot structures with the C8U substitution mutant's force-extension curves as obtained with surface-mounted optical tweezers in force-ramp mode. Unfolding occurs at a force of ~13 pN, whereas refolding is not observed upon relaxation.

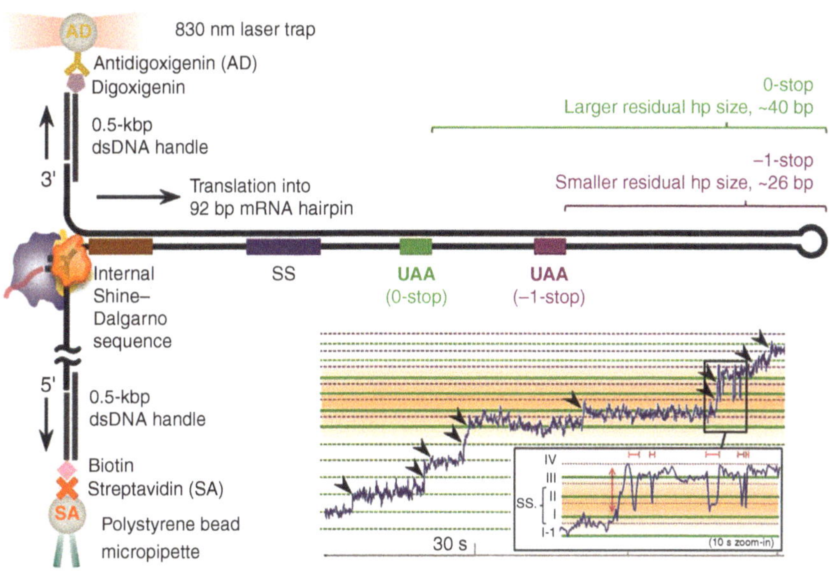

Koen Visscher, Figure 8 Single-molecule assay for detecting codon-sampling (fluctuations) and −1 frameshifting by individual ribosomes at the slippery sequence (SS). Green dotted lines are spaced one codon apart.

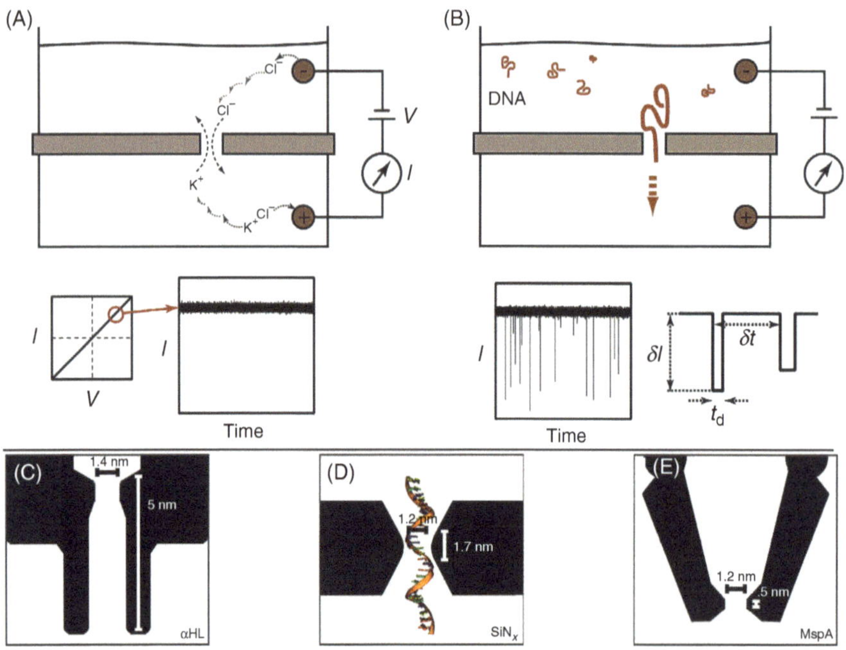

Robert Y. Henley *et al.*, Figure 1 Schematic of the nanopore ionic current detection of biomolecules,[1] as well as illustrations of the most commonly used nanopores.[2] (A) A partition containing a nanoscale aperture (nanopore) separates two chambers containing electrolyte solution. When a voltage bias is applied across the partition, a steady ionic current through the nanopore is detected (images not to scale). (B) When DNA is added to the grounded (*cis*) chamber, it is electrophoretically driven through the nanopore to the positively charged (*trans*) chamber. The passage of each individual DNA molecule results in a transient decrease in the ionic current through the nanopore. (C–E) Schematics of α-hemolysin (αHL), silicon nitride (SiN), and *Mycobacterium smegmatis* porin A (MspA) nanopores (left to right) are shown with dimension labels (SiN pores have variable geometry).

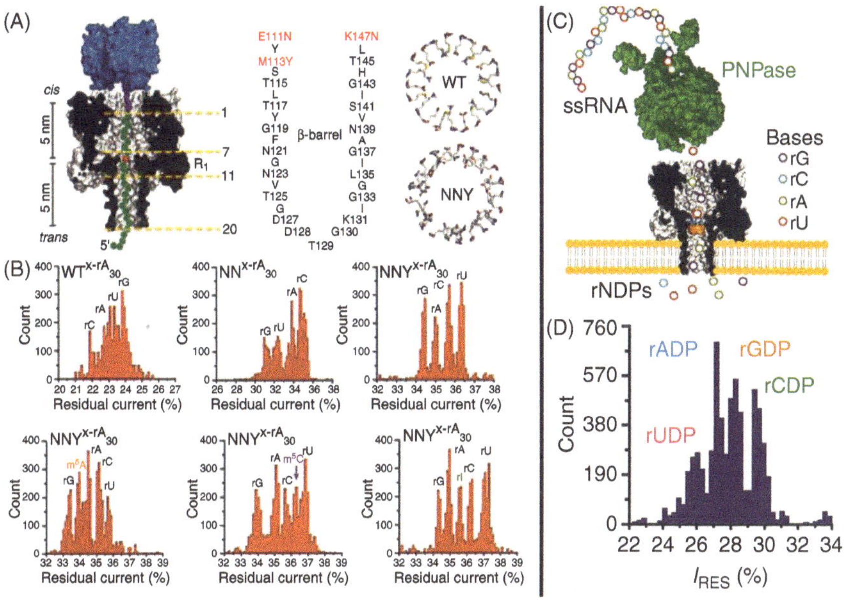

Robert Y. Henley *et al.*, Figure 2 The road toward direct RNA nanopore sequencing. (A) An illustration of the streptavidin immobilized RNA strand used by Ayub et al.[14] The four proposed constrictions are labeled, and diagrams of the WT and NNY mutant αHL pores are shown. (B) Histograms of the residual ionic current while an RNA is immobilized within the nanopore are displayed. The top row shows the superior base discrimination of the NNY mutant when compared to WT and NN pores. The bottom row demonstrates the additional epigenetic sensing ability of the NNY mutant. (C) A diagram of the proposed exonuclease-assisted sequencing method.[15] (D) Current histogram comparing the residual currents caused by binding of rADP, rGDP, rCDP, and rUDP to the molecular adapter inside the αHL β-barrel.

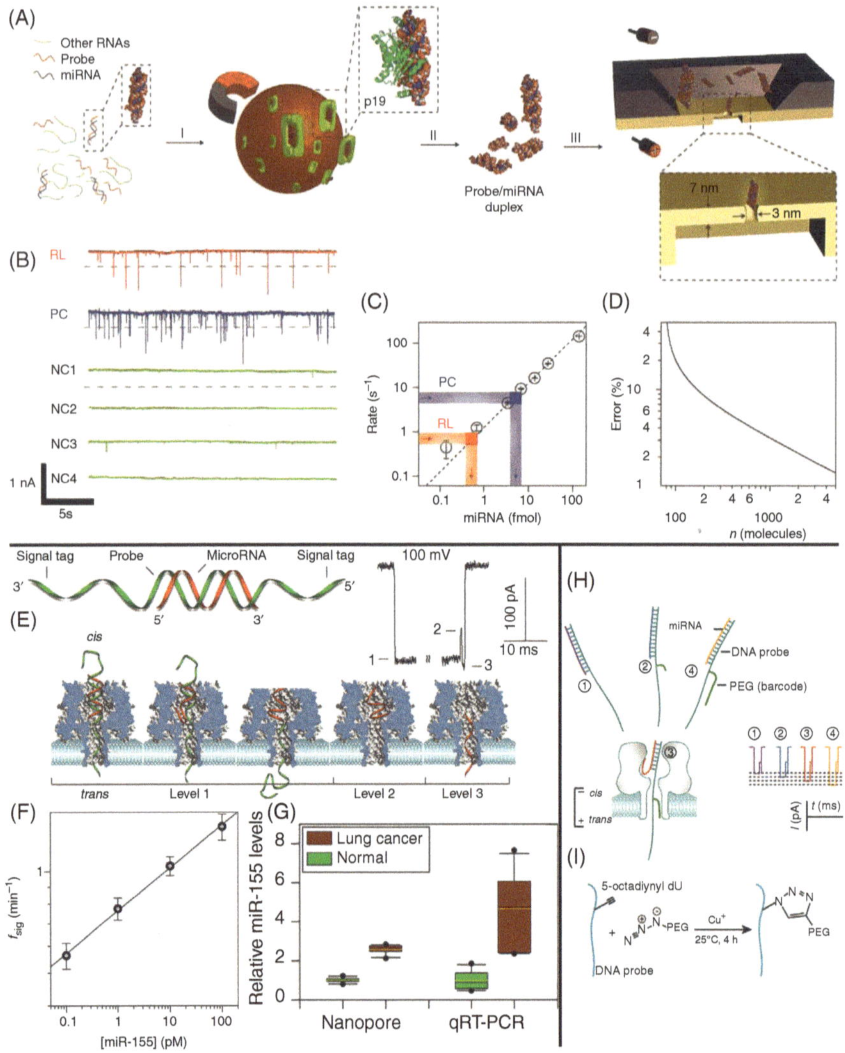

Robert Y. Henley *et al.*, Figure 3 Nanopore-based miRNA detection schemes. (A) A cartoon of the miRNA detection scheme used by the Drndić Lab.[20] Total RNA extract is incubated with probe molecules complimentary to a target miRNA sequence. This mixture is then incubated with magnetic beads that are functionalized with the p19 protein, which selectively binds to 21–23 bp dsRNA molecules. A washing step then isolates the target miRNA/probe complex, which can be readily detected using a thin SiN nanopore. (B) A sample of the ionic current traces resulting from translocations of the isolated miRNA/probe complexes. RL is the trace resulting from enrichment of miR122a, PC is a positive control, and NC refers to various negative control experiments. (C) A calibration curve is constructed by plotting the event rates as a function of miRNA concentration. The event rates of the PC and RL experiments are shown as indicated and the miRNA concentration is inferred from the calibration curve. (D)

The error in concentration as predicted from the nanopore experiments, as a function of number of translocation events. (E) A diagram of the unzipping method used by the Gu Lab.[22] A probe/target miRNA complex is shown along with the translocation process of this molecule and its resulting ionic current trace. (F) A calibration curve constructed by plotting event rate versus miRNA concentration is shown. (G) The miRNA levels as detected by the nanopore, as well as qRT-PCR are compared. (H) A schematic of the revised Gu Lab strategy,[23] employing four different polyethylene glycol (PEG) tags. (I) A diagram of the click chemistry tagging process.

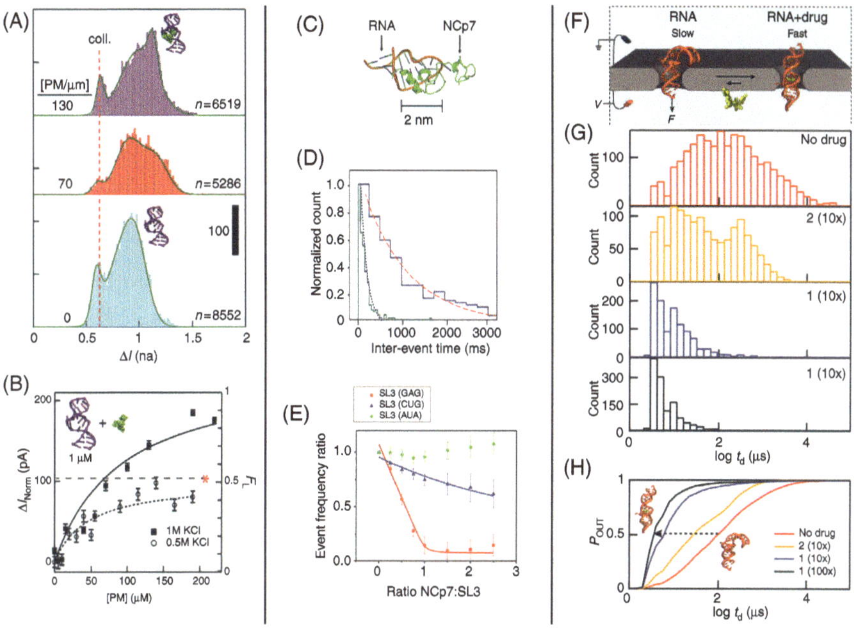

Robert Y. Henley *et al.*, Figure 4 Detection and quantification of RNA/ligand interactions. (A) Current blockade (ΔI) histograms at various ratios of paromomycin to RNA as indicated. Bottom histogram is fit to a double Gaussian and the top two are fit to triple Gaussians. A third peak emerges at large ΔI attributed to the RNA/paromomycin complex. (B) Normalized current blockade (ΔI_{norm}) is plotted versus paromomycin concentration. Right axis shows F_L values (see main text) and data is fit to Eq. (1). The dashed line and star indicate the saturated values at 0.5 M KCl. (C) Diagram of the RNA/NCp7 complex used by the Movileanu group.[35] (D) Histograms of the interevent times for RNA only (purple curve) and with added NCp7 (red curve). (E) Event frequencies normalized to RNA; only event rate are plotted against the ratios of NCp7/RNA constructs. (F) A cartoon diagram of the hepatitis C virus (HCV) conformational analysis system used by the Wanunu Lab.[36] (G) Histograms showing the log of the dwell time (t_d) for various drug concentrations as indicated. (H) Integrated normalized dwell-time distributions are plotted for various drug concentrations.

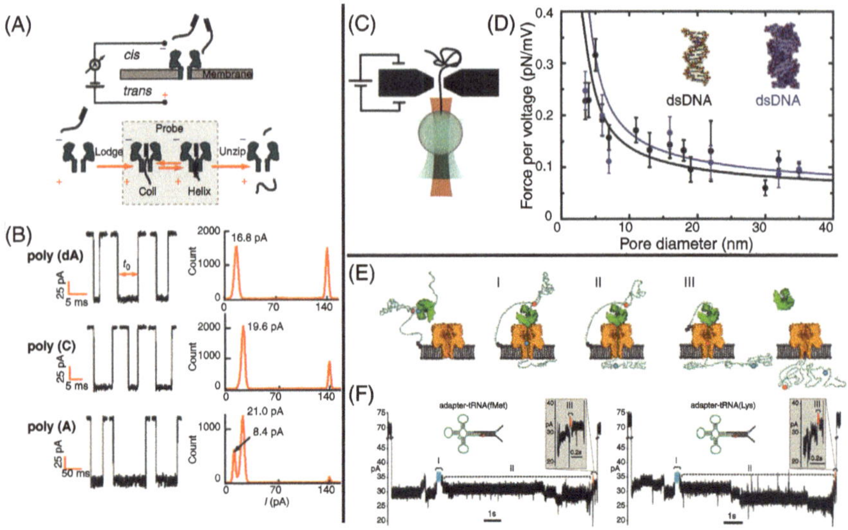

Robert Y. Henley *et al.*, Figure 5 RNA structural analysis using nanopores. (A) A schematic of the unzipping technique used to observe the helix-coil transition.[49] (B) Characteristic sample events are shown for each of the molecules studied, with each accompanied by an all-point histogram of the current data. (C) A cartoon of the hybrid optical tweezers/solid-state nanopore setup used by the Cees Dekker Lab.[51] (D) Force per voltage as a function of nanopore diameter is plotted for both dsDNA (black) and dsRNA (blue). (E) A schematic for the tRNA unfolding and transport process used by Smith et al.[52] (F) Characteristic ionic current traces observed during the capture, unfolding, and translocation of $tRNA^{fMet}$ and $tRNA^{Lys}$.

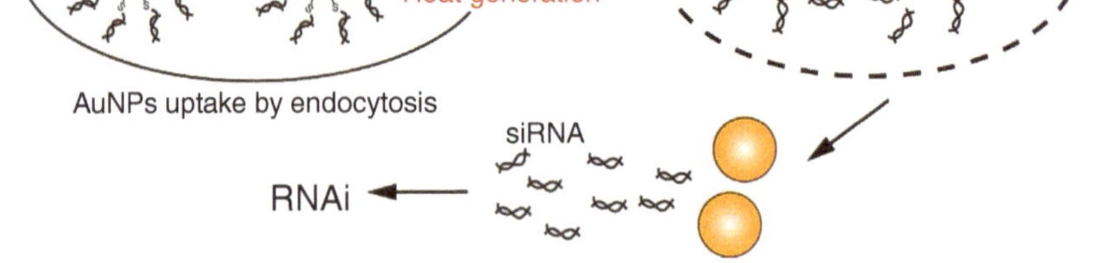

Kazunori Watanabe and Takashi Ohtsuki, Figure 1 Schematic illustration of photoinduced RNAi using AuNPs. The endosomally entrapped siRNA-conjugated AuNPs can escape from the endosomes by NIR irradiation because of NIR-induced heat generation from the AuNPs. The released siRNAs induce RNAi in the cytoplasm.

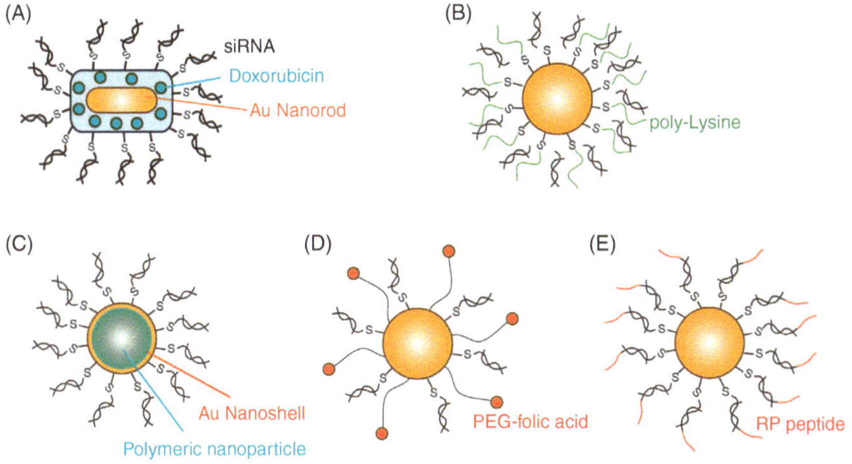

Kazunori Watanabe and Takashi Ohtsuki, Figure 2 Schematic illustrations of AuNPs. (A) Au NR with siRNA-capped mesoporous silica shell loaded with doxorubicin.[42] (B) Gold NS conjugated with poly(L-lysine) (PLL) peptides and electrostatically associated negative charged siRNAs.[44] (C) Polymer gold nanoshell (PGN) coated with pyrogallol 2-aminoethane (PAE) carrying siRNAs.[45] (D) Hollow gold nanosphere (HAuNS) carrying PEG–folic acid conjugates and siRNAs.[46] (E) Hollow gold nanoshell (HGN) coated with cancer cell-targeting peptide, RPARPAR (RP).[47]

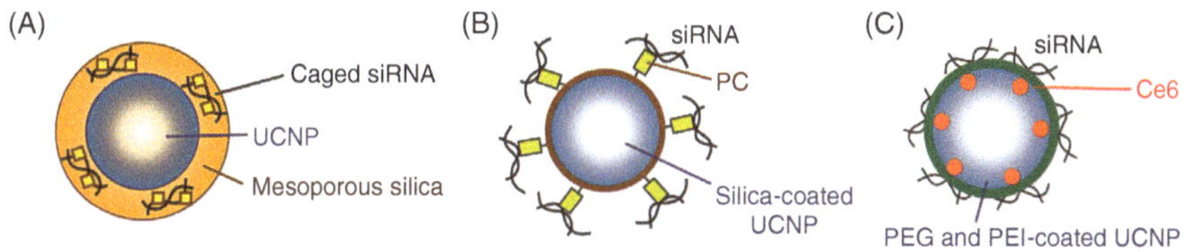

Kazunori Watanabe and Takashi Ohtsuki, Figure 4 UCNPs. (A) UCNPs coated with mesoporous silica, in which caged-siRNAs are loaded.[57,58] (B) Silica-coated UCNP carrying siRNAs via PC linker.[59] (C) PEG- and branched PEI-coated UCNPs loaded with siRNAs and chlorin e6 (Ce6).[61]

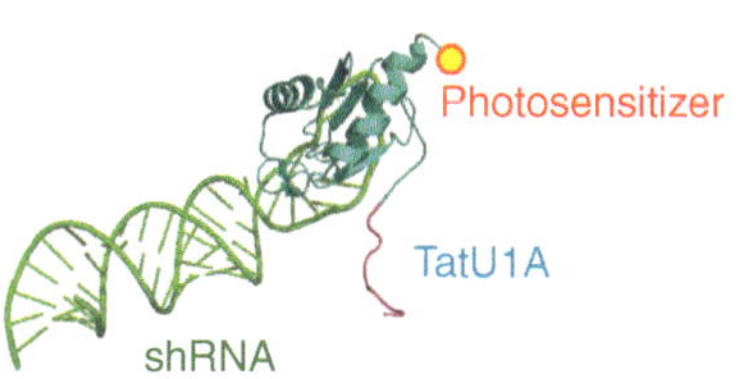

Kazunori Watanabe and Takashi Ohtsuki, Figure 5 TatU1A–PS conjugate and an shRNA. The accession numbers of the structures used here are 1R9F (siRNA) (a part of the shRNA in this figure) and 1U6B (U1A).

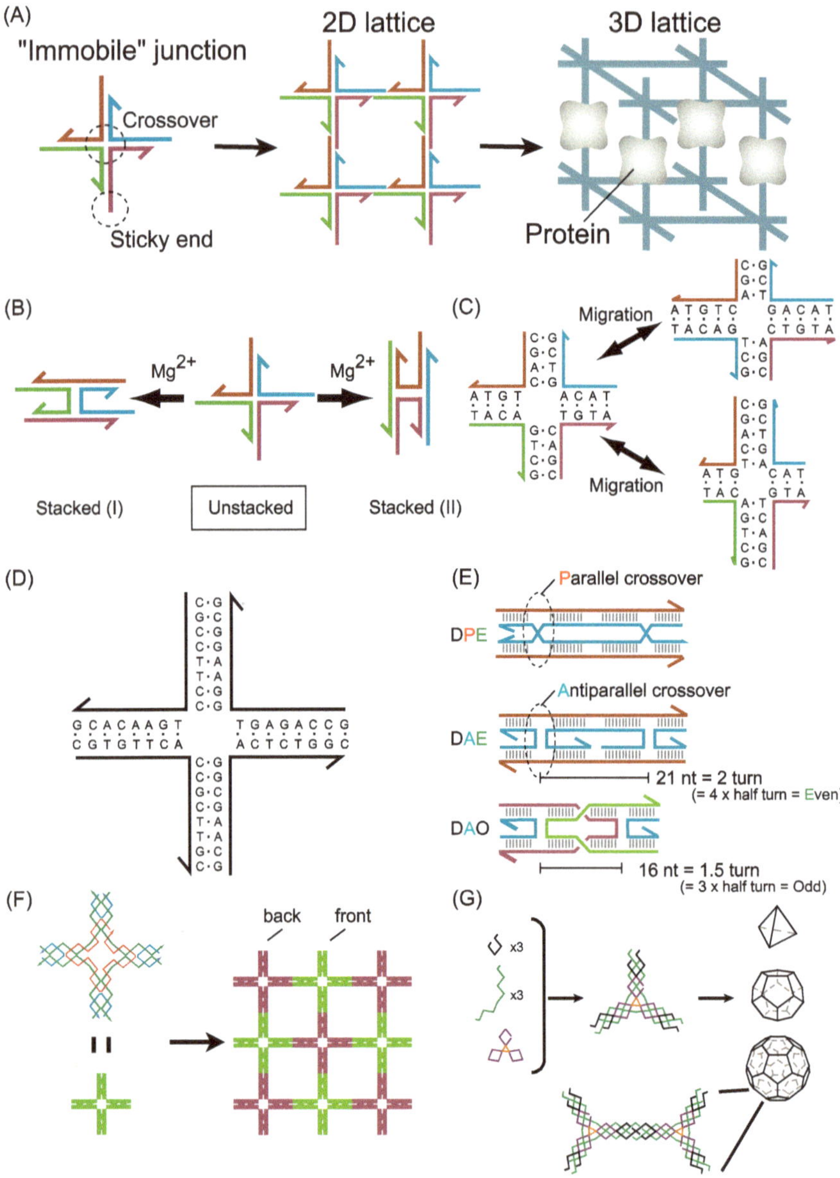

Hisashi Tadakuma *et al.*, Figure 1 Multiple-stranded tile design. (A, left) Four single-strand DNAs form Holliday junction, which has a crossover (junction) and sticky ends, acts as a unit motif of higher-dimensional structures of 2D and 3D (right). The latter could be a scaffold to organize guest proteins to facilitate protein crystallization. (B) Under low ion concentration, native Holliday junction takes planar extended conformation "unstacked form" due to the electrostatic repulsion between the negatively charged backbones. It takes "stacked form" under high ion concentration. (C) Native Holliday junction also shows migration (migrates in a random walk process). Therefore the overall structure of the Holliday junction is not unique but floppy. (D) An example of "immobile" branch point in Holliday junction obtained by means of sequence optimization. (E) In addition to "immobile" junction, "double-crossover (DX)" molecule was introduced to fix the unique structure. There are several types of DX molecule. (Top) DPE (double crossover, parallel, even number of half turns spacing), (middle) DAE (double crossover, antiparallel, even number of half turns spacing), (bottom) DAO (double crossover, antiparallel, odd number of half turns spacing), where DAE and DAO are only experimentally confirmed as stable structures. (F) Self-assembly of DX tile with periodic square cavities pattern. (G) Polyhedron and buckyball structures made by three-point-star motif method. (Part A, C, D: adapted from Seeman (1982).[32] Part E: adapted from Fu and Seeman (1993).[33] Part F: adapted from Yan et al. (2003).[34] Part G: adapted from He et al. (2008).)[35]

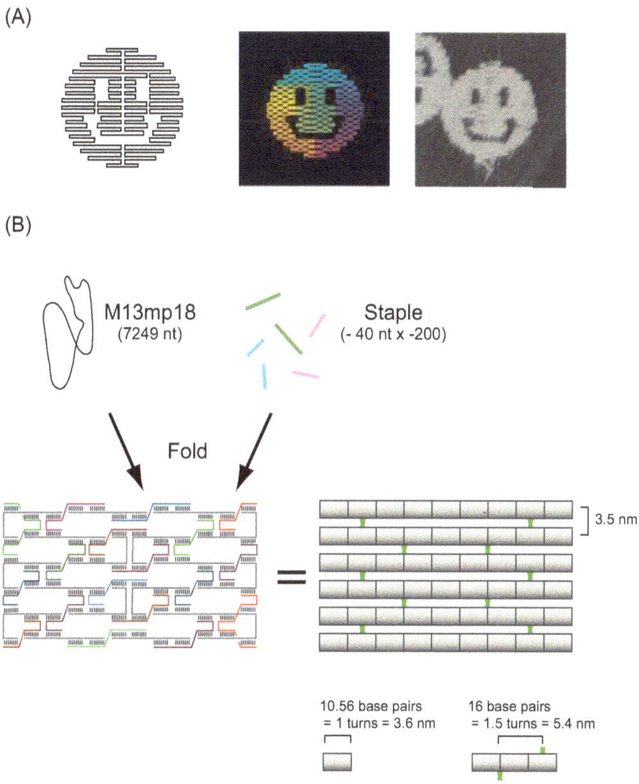

Hisashi Tadakuma *et al.*, Figure 3 Scaffold design (DNA origami). (A) Smiley face structure with DNA origami method. (B) In DNA origami method, long circular single-stranded DNA (black) is folded into the desired shape by many short single-stranded DNAs (colored), the latter typically bind to three adjacent helices, and the length is commonly 32 nucleotides, in which the central 16 nucleotides bind to one helix and the remaining two parts of 8-nucleotide ends bind to the adjacent helices. Unit pixel size is with dimensions of 3.6×3.5 nm.

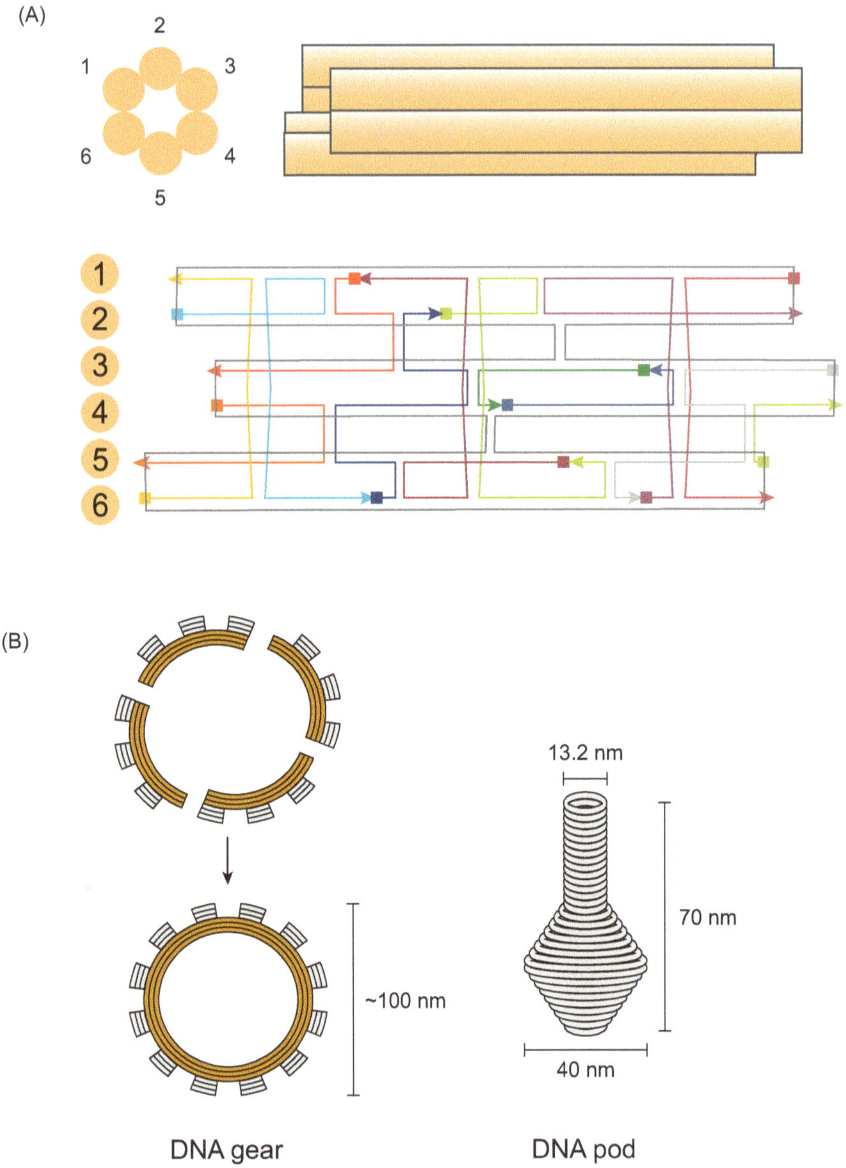

Hisashi Tadakuma *et al.*, Figure 4 3D DNA structure of Scaffold design. (A) Basic unit of scaffold designed 3D DNA structure. Cross-section (left) and side view (right) of honeycomb structure composed of six helices with sample staples using caDNAno (http://cadnano.org/, bottom).[60,70] See also Section 5.3 for detailed design processes. (B) Representative 3D structures such as gear and pod.

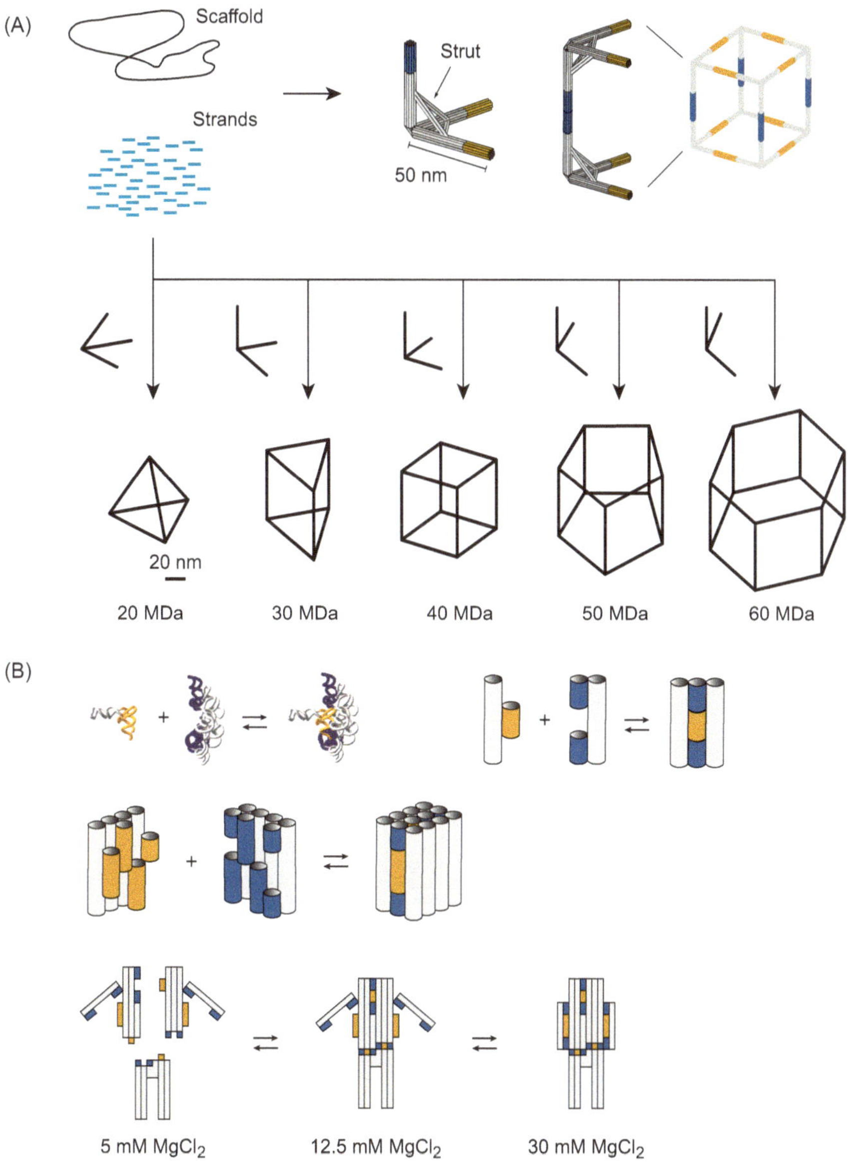

Hisashi Tadakuma *et al.*, Figure 5 Higher assembled DNA structure. (A) Pole and joint approach to construct higher assembled DNA structure. A tripod is composed of one set of DNA origami structure and used as a basic unit that has three arms and binds with each other at the apical point. The structure of the end product is defined by the angle between the arms. (B) Shape-complementarity method to construct higher assembled DNA structure. (Top) DNA structure binds to its paired structure (top right) in a way that RNase P recognizes its substrate tRNA (top left). (Middle) Upon two motifs engagement, nucleobase stacking interactions occur at the double helical interface of the complementary shapes, but only upon correct fit of the helices. (Bottom) Shape-complementary interaction is highly affected by ion concentration, therefore higher assembled DNA structure, for example, nanorobot of 15 MDa, can be reversibly transformed in three different conformation states: disassembled, assembled with open arms, and assembled with closed arms, respectively, by changing the Mg^{2+} concentration.

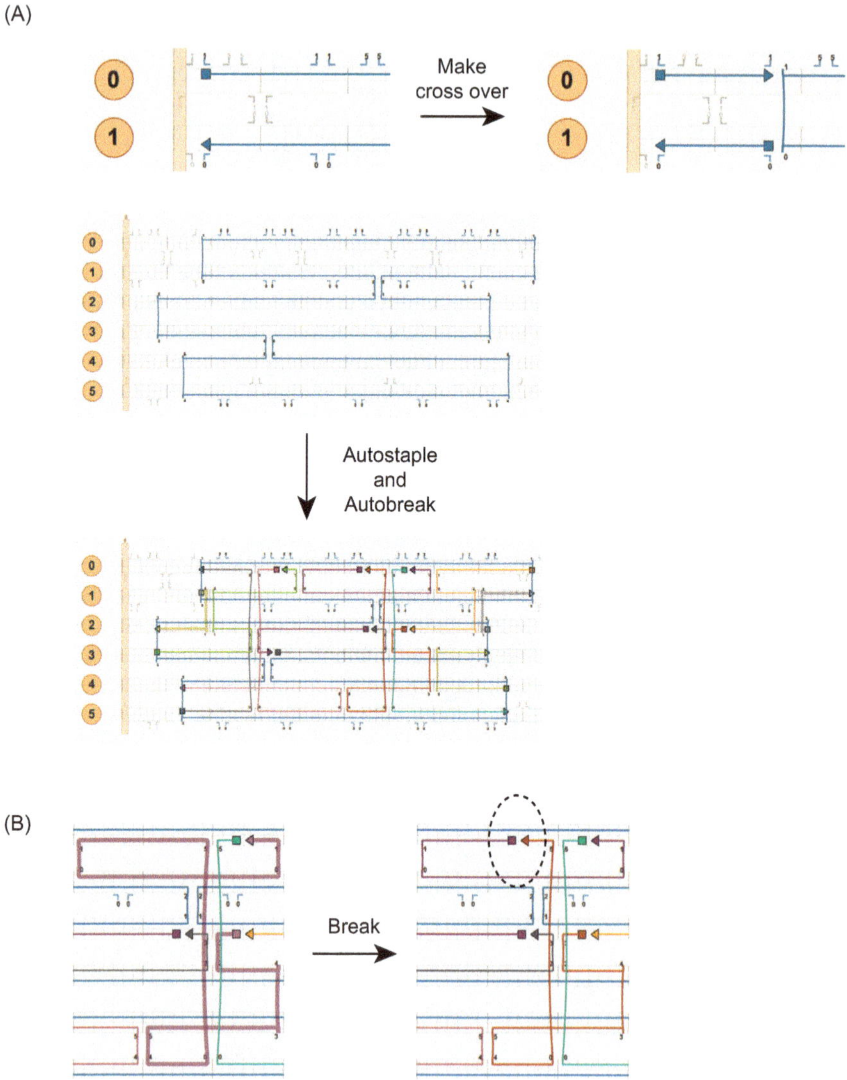

Hisashi Tadakuma *et al.*, Figure 8 Design software assisted DNA nanostructure design. (A) Captured images of caDNAno (http://cadnano.org/), an open-source computer-aided design (CAD) software package with a graphical user interface, developed by

William Shih and Shown Douglass' group.[70] By clicking the blue mark on the vicinity of the line, crossovers can be introduced at the predicted potential sites. After drawing the scaffold design, "Autostaple" and "Autobreak" functions fulfill the suggested staples. (B) With manual adjustment, final staple sequences are determined.

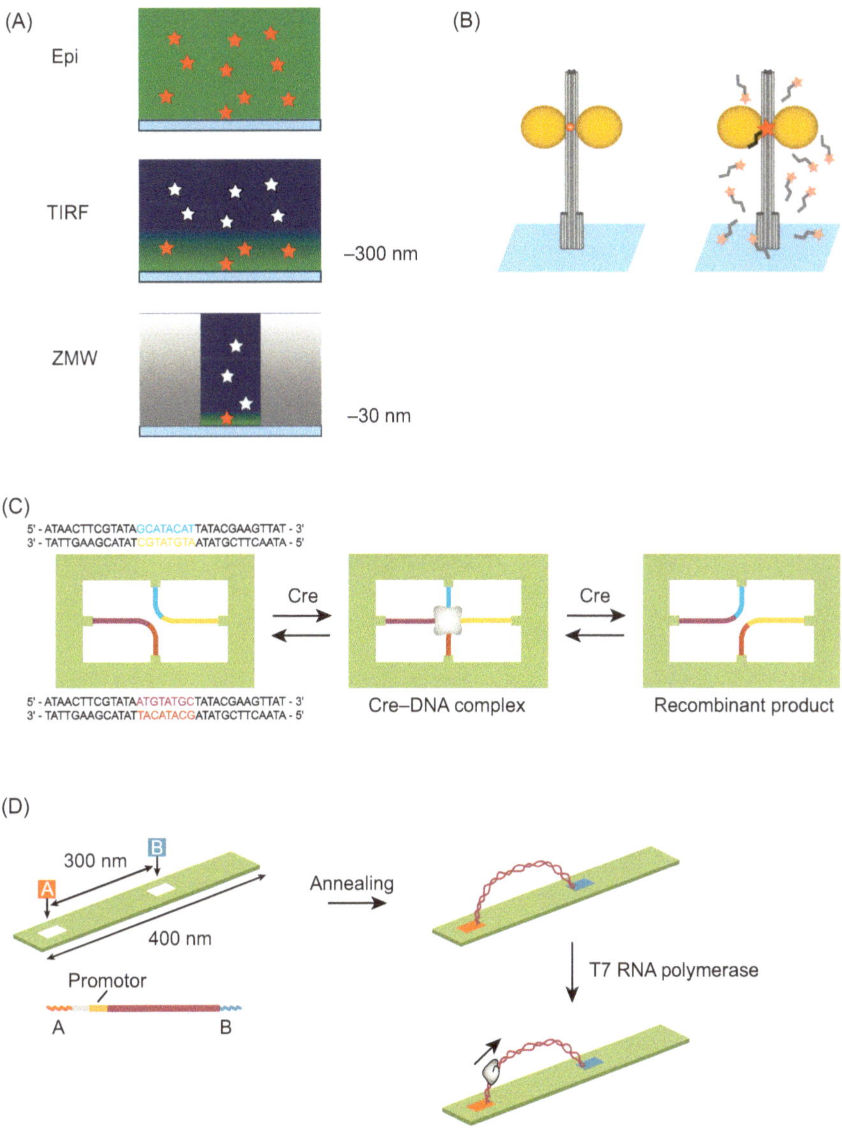

Hisashi Tadakuma *et al.*, Figure 9 Imaging of RNA using DNA nanostructures. (A) Confinement of the excitation volume is the key for single-molecule imaging with high signal to noise ratio. Epi: epi-fluorescence imaging, where all fluorescent dyes are excited, making high background noise in observing single molecule anchored on the glass surface (light blue block). TIRF: total internal reflection fluorescence microscopy reduces the excitation volume in the depth direction, illuminating vertical (*z*) direction of up to 300 nm, reducing the background noise and enabling to observe single enzyme reaction in solution.[6] ZMW: in addition to *z*-direction, *x*–*y* horizontal direction is also confined (~100 nm), achieving zeptoliter sampling volumes and single-molecule imaging under the presence of micromolar background fluorescent dye.[139] For details see chapter: Probing the Translation Dynamics of Ribosomes Using Zero-Mode Waveguides written by Tsai et al. (B) Nanoantenna has two 100 nm gold nanoparticles on a pillar-shaped DNA origami of 220 nm height, achieving single-molecule imaging in-between the gold particles. (C) Real time visualization of Cre recombinase reaction on DNA origami achieved by high-speed atomic force microscopy (HS-AFM). (D) Sliding of T7-RNA polymerase (T7-RNAP) on a template gene embedded into DNA origami was also visualized by HS-AFM.

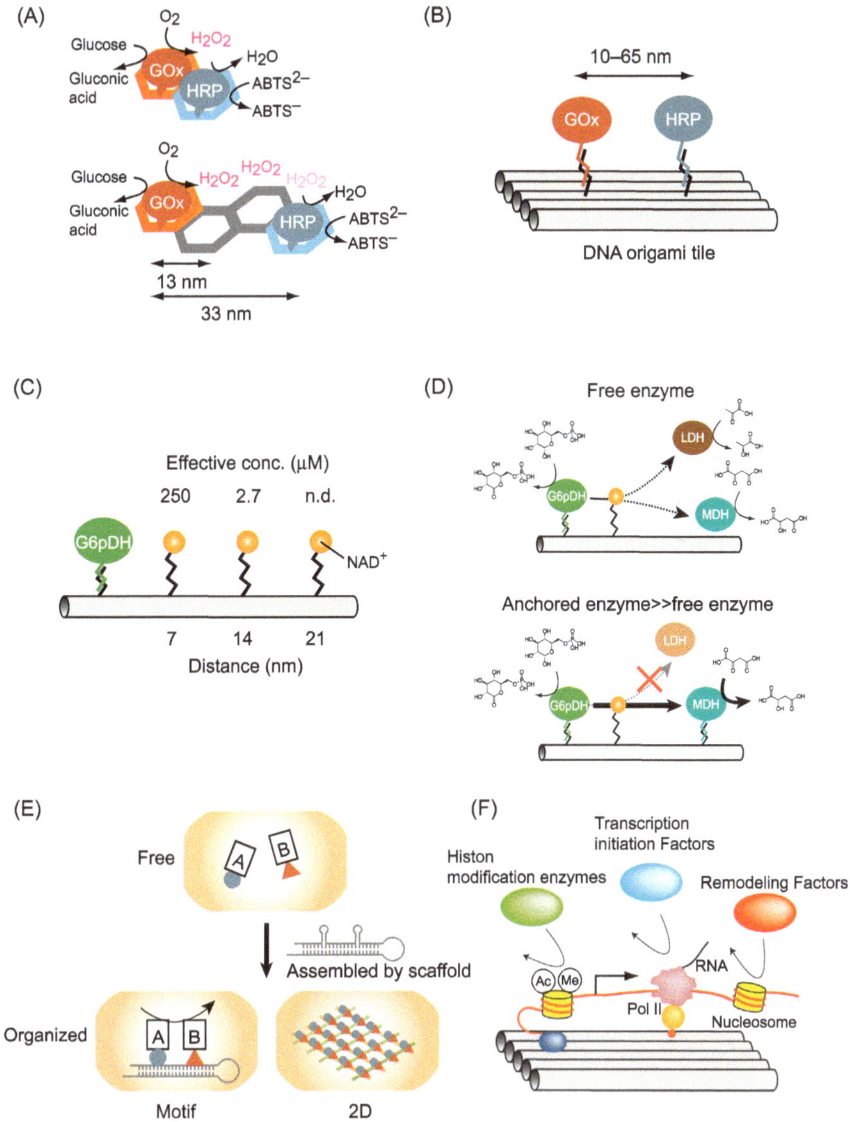

Hisashi Tadakuma *et al.*, Figure 10 Controlling biochemical reaction by molecular layout design. (A) Cascade enzymatic reaction is enhanced by organizing the intermolecular distance of enzymes, for example, GOx: glucose oxidase and HRP: horseradish peroxidase. (B) Using DNA origami, intermolecular distance effect was examined systematically. (C) Evaluation of the effects of the distance between cofactor NAD^+ and the enzyme (G6pDH). The NAD^+ molecule anchored 7 nm apart from the enzyme is comparable to the effective concentration of 250 μM. When anchored at 14 nm, it is steeply decreased to 2.7 μM. (D) The strong dependence on the distance of effective concentration used to design controlled enzymatic pathway. The product of the first enzymatic reaction, NADH, is mainly consumed by the tethered second enzyme (MDH: malic dehydrogenase), but not by the freely diffusing competitor enzyme (LDH: lactate dehydrogenase). (E) In vivo enhancement of cascade enzymatic reaction using RNA scaffold and RNA motif binding proteins. (F) Gene nanochip, integrating enzyme (RNAP), and substrate (template gene) on the DNA origami, may allow to evaluate the precise mechanism of the multiple cascade enzymatic gene expression reaction in detail.

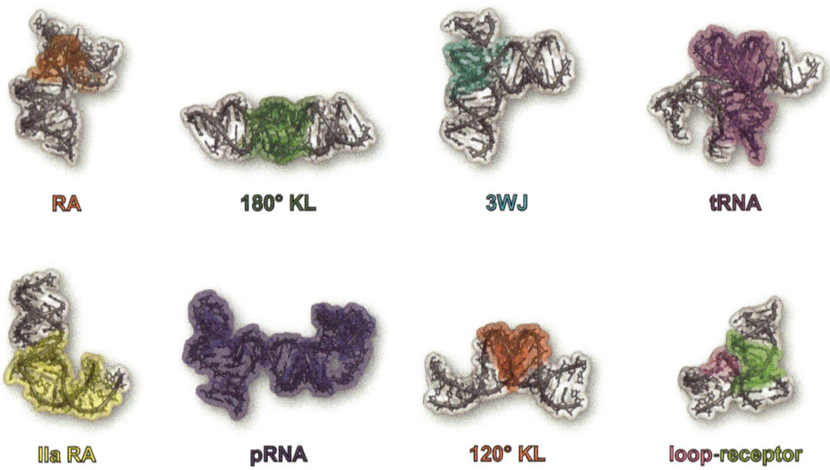

Hirohisa Ohno and Hirohide Saito, Figure 1 RNA structural motifs. Various RNA structural motifs have been found out from natural RNA molecules. Each 3D data was obtained from PDB (id is shown later). Orange, right-angle *(RA)* motif (extracted from 1JJ2[8]); green, 180° kissing-loop *(KL)* motif (1JJM[9]); light blue, three-way junction *(3WJ)* motif (extracted from 4V4Q[10]); purple, *tRNA* motif (extracted from 4TNA[11]); yellow, *IIa RA* motif (2NOK[12]); blue, *pRNA* motif (3R4F[13]); red, 120° *KL* motif (2BJ2[14]); pink and light green, *loop-receptor* motif (extracted from 1HR2[15]). Core region of each motif is colored and white region corresponds to normal Watson–Crick type double-stranded helix region. All of them are drawn in the same scale.

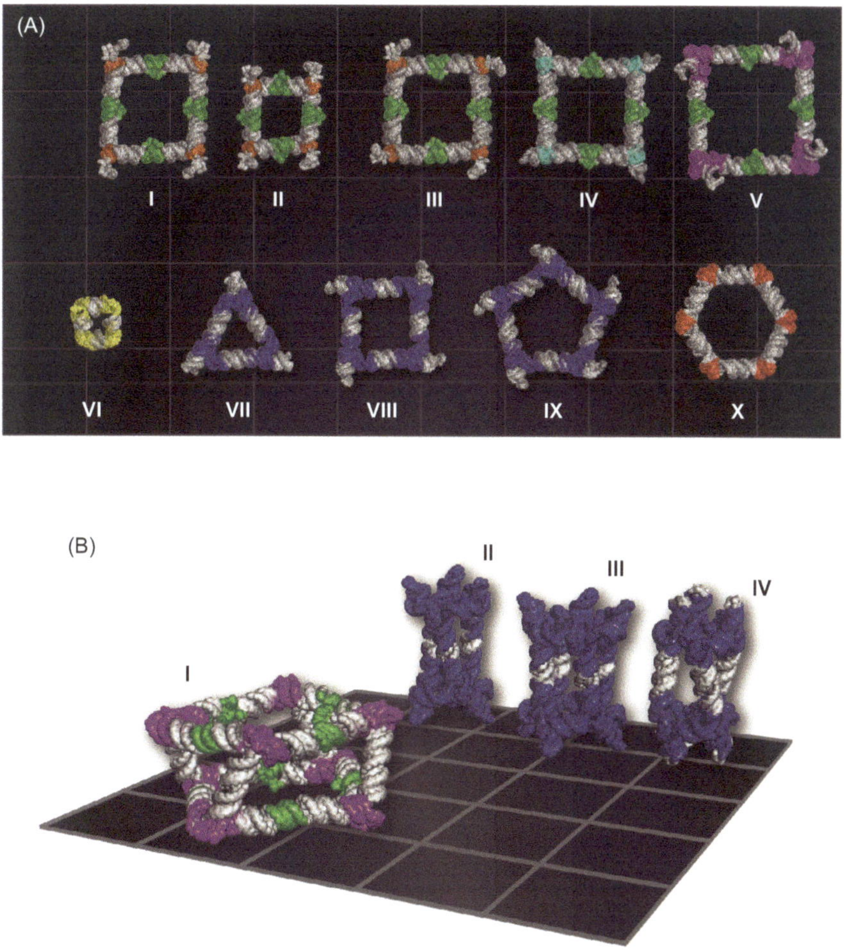

Hirohisa Ohno and Hirohide Saito, Figure 2 RNA nanostructures. (A) Various 2D and (B) 3D nanostructures have been built with RNA. The 3D models of IIa RA square is obtained from PDB (id: 3P59[17]). Other models are built according to each reference. Colored regions indicate motif regions and their colors are corresponding to Fig. 1. The mesh size of grid is 10 × 10 nm.

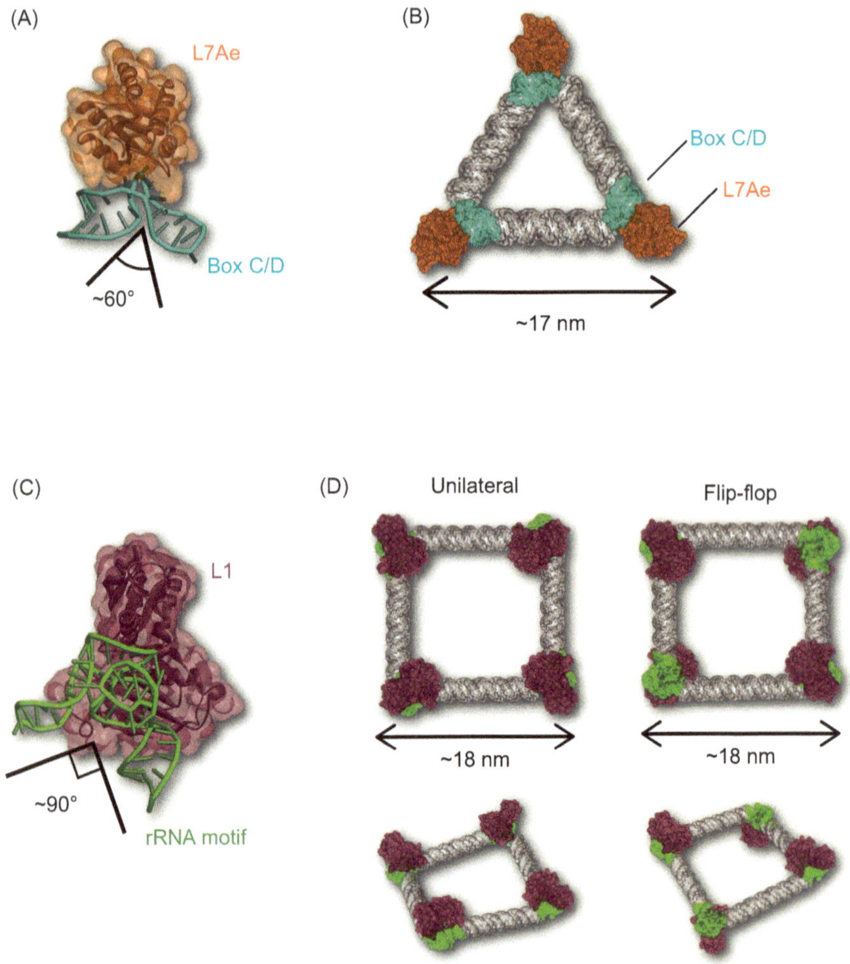

Hirohisa Ohno and Hirohide Saito, Figure 3 RNP nanostructures. (A) Box C/D-L7Ae module. Box C/D RNA (light blue) is bent and fixed at ~60 degrees by the bound L7Ae protein (orange); (B) 3D model of L7-triangle. Colored region at apex corresponds to RNP module. White regions consist of ordinary double-stranded RNA; (C) rRNA motif-L1 module. The rRNA motif (green) forms the right angle shape in the complex with L1 protein (purple); and (D) 3D models of two-types of squares.

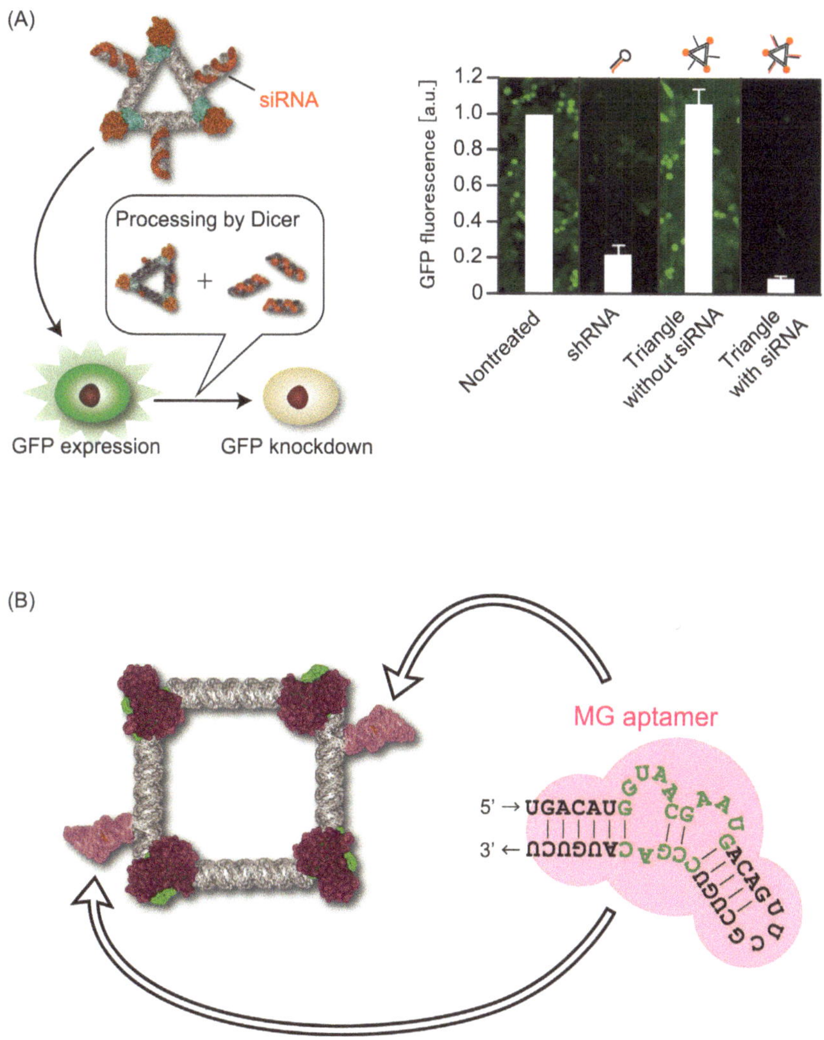

Hirohisa Ohno and Hirohide Saito, Figure 4 Introduction of functional RNA into RNP nanostructure. (A) Anti-GFP siRNA was installed onto the line regions of triangle (left). The siRNA-carrying triangle shows the effective suppression of target GFP gene expression (right graph; redrawn from ref. 31); and (B) 3D model of MG aptamer (pink)-installed L1-square (left). The 2D structure of inserted aptamer is shown (right). The core region is indicated by green letters.

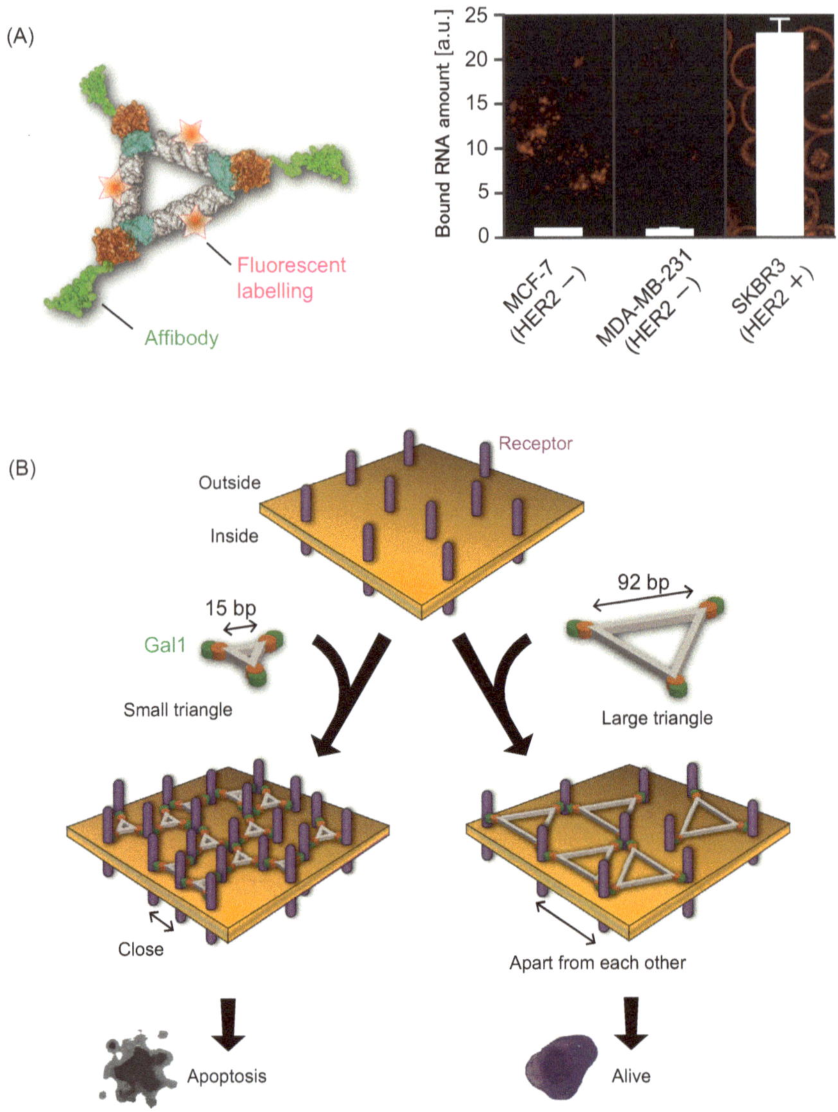

Hirohisa Ohno and Hirohide Saito, Figure 5 Functionalization of RNP nanostructure by the introduction of protein. (A) For cell type-specific recognition, affibody (green) was introduced onto the L7-triangle (left 3D model). Triangular RNA was labelled by fluorescent dye to detect the binding to cell surface (pink star). The affibody-carrying triangle can detect specifically HER2-expressing cell (right graph; redrawn from ref. 31). (B) Control of intermolecular distance by using RNP nanostructure. Based on their size, triangular RNPs were able to control the assembling of receptor proteins on the cell-surface.

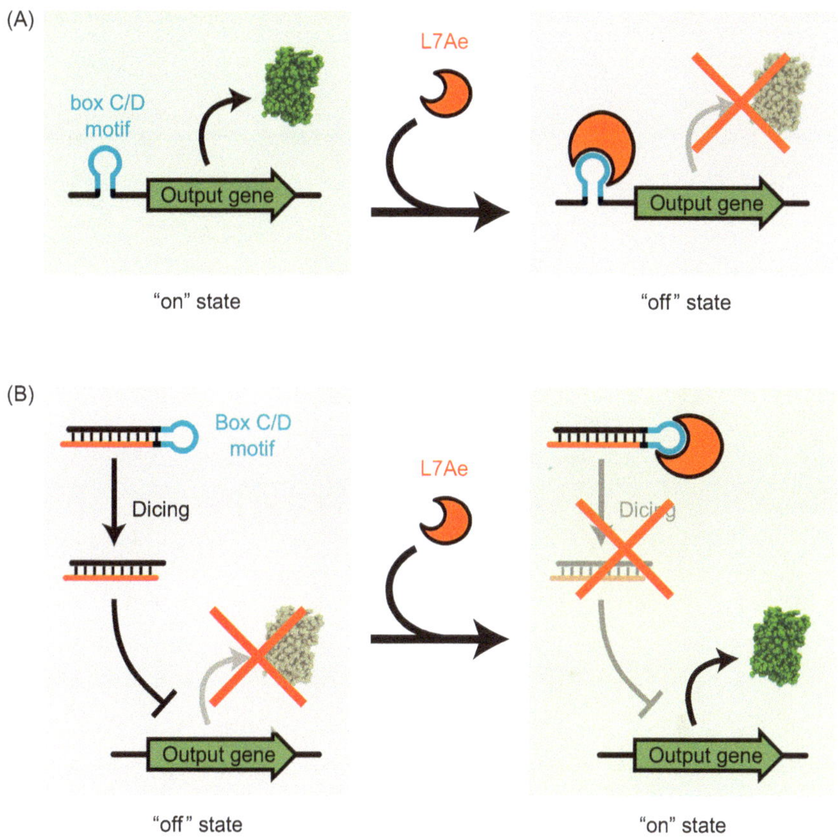

Hirohisa Ohno and Hirohide Saito, Figure 6 RNP switch to control gene expression. (A) "OFF switch" by translational inhibition. In the absence of L7Ae protein, the mRNA can be translated by ribosome and the output gene is expressed ("on" state; left). However, in the presence of L7Ae, it binds to box C/D motif on the mRNA and impedes the translation ("off" state; right). (B) "ON switch" employing RNA interference pathway. Here, RNP motif is inserted to a premicro RNA (pre-miRNA). In the absence of L7Ae protein, the pre-miRNA is normally processed into micro RNA (miRNA) by Dicer. The miRNA represses the expression of reporter mRNA ("off" state; left). In contrast, L7Ae binds to pre-miRNA and impedes the processing by Dicer in the presence of L7Ae. Therefore, reporter mRNA is expressed ("on" state; right).

FSC
www.fsc.org
MIX
Papier | Fördert gute Waldnutzung
FSC® C083411

Kolibri 360 GmbH
Ferdinand-Jühlke-Straße 7
99095 Erfurt, Deutschland
produktsicherheit@kolibri360.de